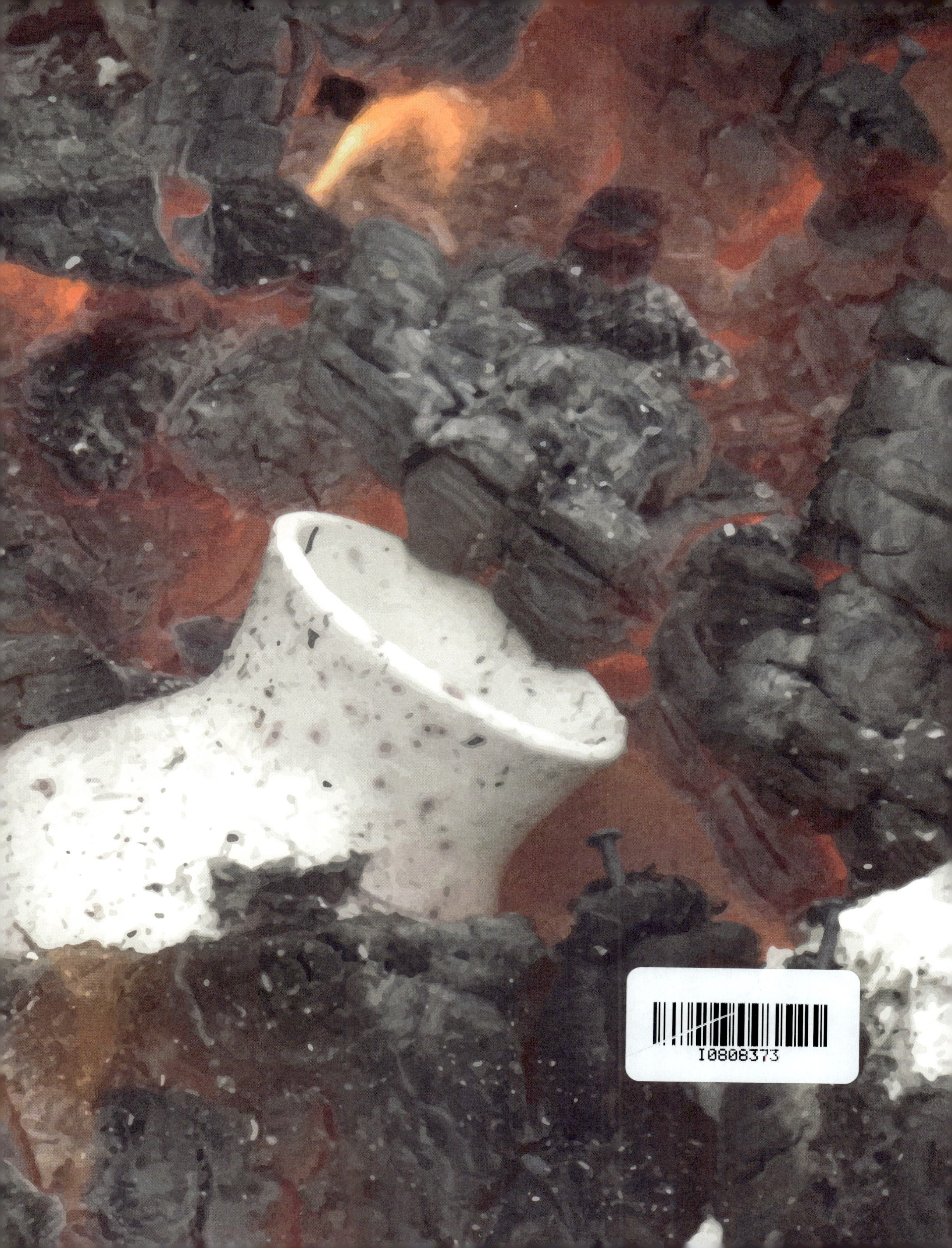

I0808373

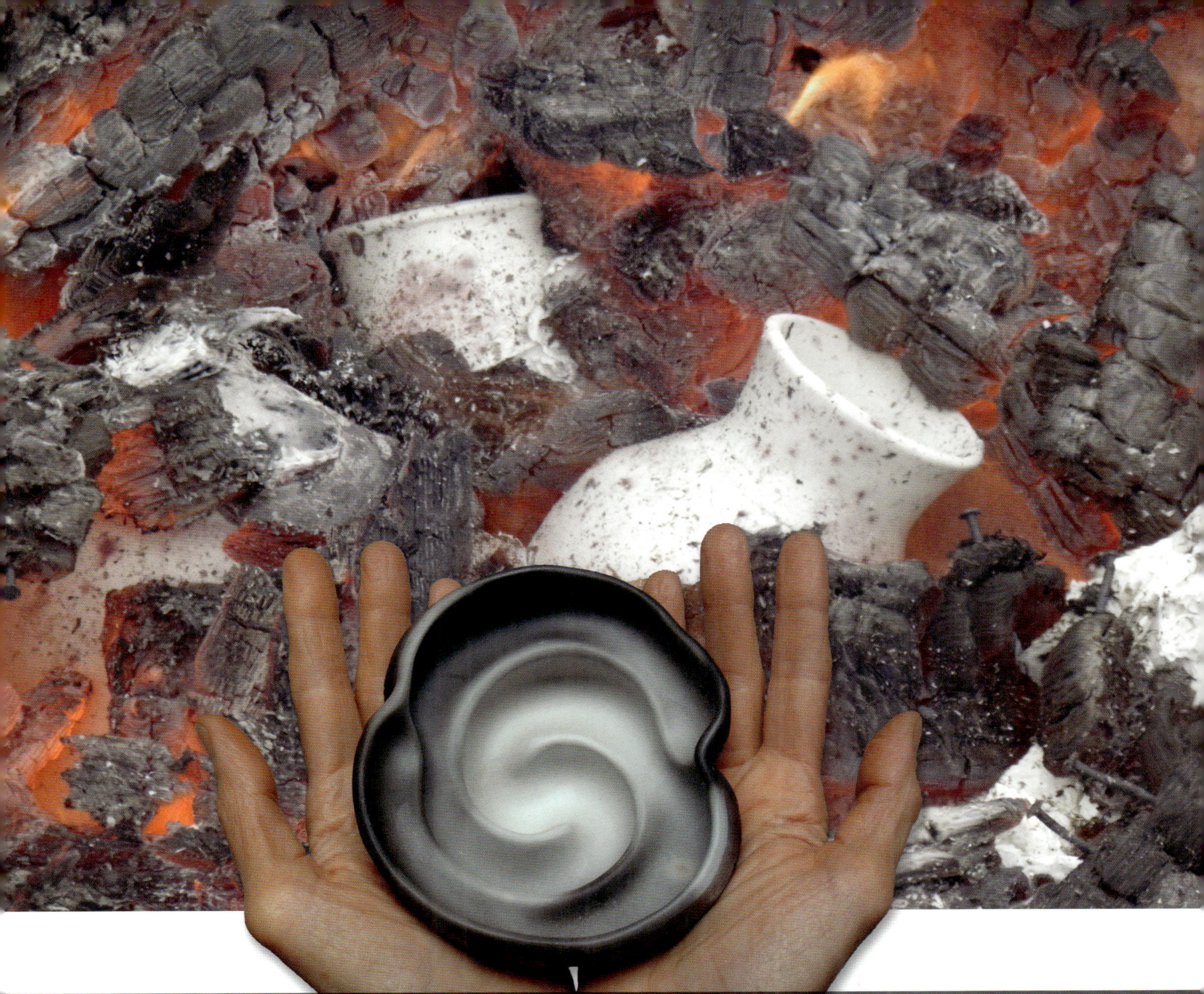

PIT FIRING CERAMICS

MODERN METHODS, ANCIENT TRADITIONS

DAWN WHITEHAND

Schiffer Publishing Ltd

4880 Lower Valley Road • Atglen, PA 19310

Other Schiffer Books on Related Subjects:
Ceramics Today. Jeffrey B. Snyder, ed.
ISBN: 978-0-7643-3465-8. $50.00
Wood-Fired Ceramics: 100 Contemporary Artists.
Amedeo Salamoni. ISBN: 978-0-7643-4533-3. $59.99

Copyright © 2013 by Dawn Whitehand

Library of Congress Control Number: 2013944977

All rights reserved. No part of this work may be reproduced or used in any form or by any means—graphic, electronic, or mechanical, including photocopying or information storage and retrieval systems—without written permission from the publisher.

The scanning, uploading and distribution of this book or any part thereof via the Internet or via any other means without the permission of the publisher is illegal and punishable by law. Please purchase only authorized editions and do not participate in or encourage the electronic piracy of copyrighted materials.

"Schiffer," "Schiffer Publishing, Ltd. & Design," and the "Design of pen and inkwell" are registered trademarks of Schiffer Publishing, Ltd.

Designed by Justin Wakinson Cover by Bruce Waters
Type set in Agency FB/Dutch 809 BT

ISBN: 978-0-7643-4172-4
Printed in the United States

Published by Schiffer Publishing, Ltd.
4880 Lower Valley Road
Atglen, PA 19310
Phone: (610) 593-1777; Fax: (610) 593-2002
E-mail: Info@schifferbooks.com

For the largest selection of fine reference books on this and related subjects, please visit our website at **www.schifferbooks.com.** You may also write for a free catalog.

This book may be purchased from the publisher. Please try your bookstore first.

We are always looking for people to write books on new and related subjects. If you have an idea for a book, please contact us at proposals@schifferbooks.com

Schiffer Books are available at special discounts for bulk purchases for sales promotions or premiums. Special editions, including personalized covers, corporate imprints, and excerpts can be created in large quantities for special needs. For more information contact the publisher.

CONTENTS

LIST OF ILLUSTRATIONS

PREFACE

While researching pit firing for my PhD, I experienced difficulty finding relevant information. Most books, whether they approached ceramics from a historical or art/craft stance, took a chronological and descriptive approach to ceramics rather than exploring methods of production, such as firing techniques. Information that I was able to source was often in specialist literature such as archaeological or anthropological journals and books, however such publications are not readily available to artists, ceramicists, historians, or others interested in ceramics and/or firing methods of ceramics. This book is an attempt, therefore, to draw together information from a variety of sources and present it in an informative and accessible manner.

A wide range of sources was researched during the development of this book, including archaeological, anthropological, historical, and sources on the arts, along with actual artists and ceramicists themselves. Despite this diversity, conclusions about the beginnings of pottery and the first firing structures remain uncertain. While researching firing structures in the Indus region, for example, Heather Miller admits that it is often difficult to be sure about what was fired in excavated structures, especially earlier sites where excavators did not record information about wasters found in or around the firing structures. Also, there was often ambiguity concerning the purpose of some structures: whether they were used for cooking, as ritual fire altars, or as fire pits (Miller 1997, 45-6). Ultimately pits used for firing are ephemeral and leave few surface indicators behind, especially as time goes by and the integrity of an archaeological site lessens. Such research problems will be discussed in more detail in chapter one.

Chapter one examines the history of pit firing, and the various theories that have been proposed by historians, archaeologists, and ceramicists. Since the origin of the discovery of ceramics is lost in the shadows of time, many of these theories remain just that: theory.

Chapter two explores the history of contemporary pit firing, and how it came to be reinvented and incorporated into the modern ceramicist's oeuvre. The continued use of pit firing in traditional indigenous societies is discussed, and the rediscovery of pit firing as an artistic technique in modern studio practice is explored through the prism of Australia and the United States of America (USA).

Chapter three, a "how to" section, explores the working methods of a selection of contemporary ceramic artists, and includes a series of photo essays illustrating the different pit firing methods employed by a multitude of contemporary ceramicists. Also examined are health and safety issues pertinent to pit firing.

Chapter four is a section that profiles a selection of contemporary ceramicists from around the world that incorporate pit firing into their art practice. Each profile includes images of that ceramicist's artworks and the rationale behind why they choose this organic and ancient firing method.

Chapter five is the conclusion and discusses what characteristics of pit firing appeal, most commonly, to the ceramicists in this book. Pit firing's popularity as a method of ceramic firing in the twenty-first century, and the reasons for its continued appeal, will also be considered.

Finally, a glossary and bibliography are included so that readers will have easy access to technical terms and will be able to research any aspects covered in this book, should they so wish.

While this book is not a conclusive overview of the ceramicists that currently include pit firing in their studio practice, it does provide a diverse range of approaches and outcomes. Artists included come from Australia, England, Austria, Canada, New Zealand, and the USA. Thank you to all the ceramicists who have generously donated images of their work and statements about what they do and why. Without this information the book could not have come into being.

CHAPTER ONE
PREHISTORY AND EARLY HISTORY

> Compared to the use of energy in many marginal industrial processes and the profligate waste of fuel for needless lighting and space heating, the demands of potters are modest indeed, and are justified on the grounds of practical and esthetic utility. And kilns, even poorly constructed ones, are quite efficient in that they retain heat in a closed space where it can do its work. (Rhodes 1981, viii.)

WHAT IS A KILN?

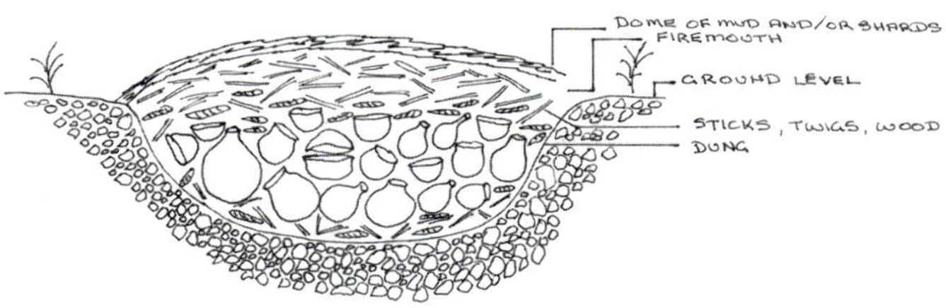

Pit Kiln. Diagram: The Author

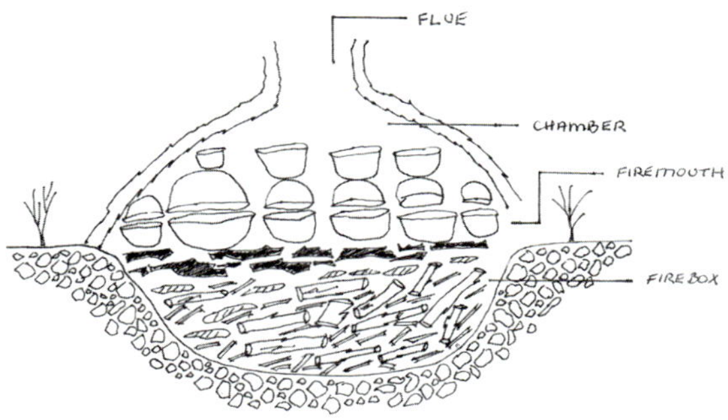

Simple updraught kiln, showing the basic elements of flue, chamber, firemouth, and firebox. The wares are separated from the fuel by a layer of ceramic shards. *Diagram: The Author*

When defining a kiln, we may consider two aspects: its appearance and its purpose. At its most basic, appearance-wise a kiln is made up of the firebox, the fire mouth, the chamber, and the flue. These physical properties mean that the ceramic wares are within an enclosed space and are separate from the fuel. There is much variation in contemporary times as technology has evolved and gas and electric kilns have developed, but in essence the purpose has remained the same: to hold ceramic wares, generate heat, and transfer that heat to the ceramic wares (Rice 1997, 253). Firing without the physical properties of a kiln is referred to as an open or bonfire firing. Pit firing, however, is conducted within an enclosure, but does not exhibit the formal qualities of a kiln. Prudence Rice, therefore, considers pit kilns functionally intermediate between open firings and simple updraught kilns in that they are within an enclosure (Rice 1997, 254). Owen Rye refers to pit firings as a mixed firing (Rye 1976, 165) because there is no separation of the ceramic wares from the fuel.

PREHISTORIC KILN RESEARCH

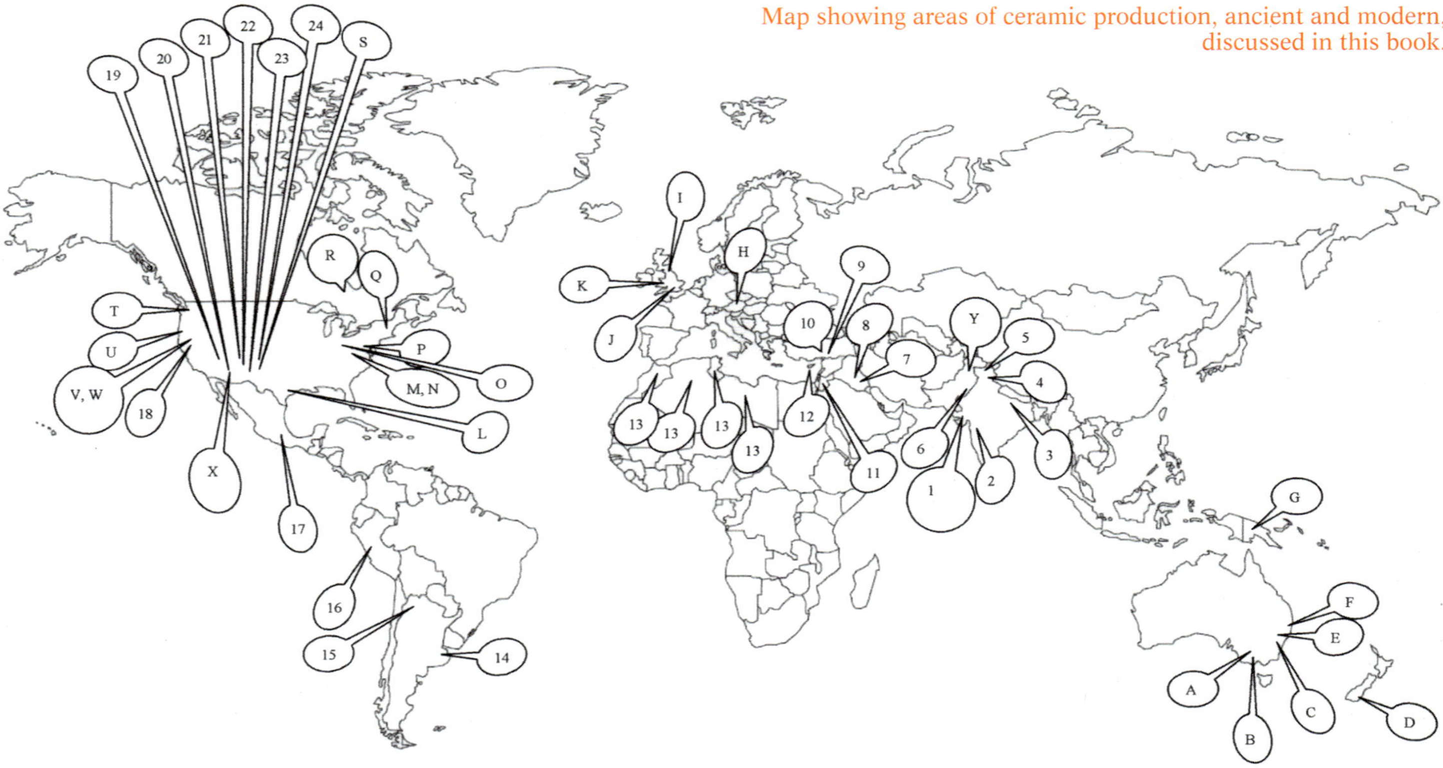

Map showing areas of ceramic production, ancient and modern, discussed in this book.

English anarchist poet, and critic of literature and art, Sir Herbert Read (1893–1968), claimed pottery to be a fundamental art form, tied to the needs of community, thereby making it a cornerstone by which to judge a society (Clark & Hughto 1979, 1), and, indeed, until recently, much archaeological research was devoted to the stylistic and iconographical study of ceramic shards, which were used as an archaeological and anthropological barometer, measuring a society's sophistication on social, religious, and technological levels. Ceramic shards make up the greatest percentage of remnants unearthed from prehistoric sites, and are invaluable in providing cultural information, such as customs, rituals, and belief-systems, about the societies that designed and produced them.

It was widely believed that the type of ceramic relic unearthed from the ground measured the history of humankind, its development and evolution from a nomadic to an agriculturally based civilization. This transition was thought to be accompanied by a shift from symbolic ceramic objects to items of a more functional and utilitarian nature. Generally, these earlier studies attempted to dissect manufacturing and developmental processes from the actual artifact. As noted by archaeologist Izumi Shimada, this approach is inherently flawed by the very nature of the artifact, which was often looted from tombs, removed from its context, relocated in museums, and often represented a technologically and artistically superior example of the culture being examined (Shimada 1997, 104). Early archaeological expeditions therefore desacralized these objects, thus neutralizing their meaning. This early research was often object based, and was used to build a system of classification and authenticity, rather than produce insights into the culture. By definition ceramic is hardened by fire, rendering it permanent; without this heat treatment, it is merely clay. Wherever ceramic shards are found, therefore, firing structures—simple or complex—must have existed.

Though production sites had been unearthed, there was little interest in kilns or the manufacturing process. As pointed out by archaeologist turned potter, Owen Rye, "[s]cientific awareness can become cultural bias" because it "will show us how *we* see past cultures, not how the people saw themselves" (Rye 1981, 3).

The limitations of archaeological methodologies in the past and the ephemeral nature of pit firing, which seldom leaves any surface indicators, has often led to previous research not recognizing some ancient firing sites. Often pit kilns, after many centuries, especially if small and only used briefly, can be almost impossible to recognize in an archaeological surface survey. Rye, during his field studies in Pakistan, where he researched traditional pottery techniques, visited a potter in the Chitral Valley who fires in a shallow, circular depression. After the firing, he removes the wood-ash to use as fertilizer in his fields, thereby leaving very little evidence for future archaeologists (Rye 1976, 165).

Alternatively, Gary Feinman and Andrew Balansky, in their examination of firing in Oaxaca, point to what they call the "tyranny of ethnographic analogy" (Feinman 1997, 141) as a contributing reason for the lack of pit kiln findings in Mesoamerica. Because this region was a high producer of ceramics, archaeologists have consistently looked for a more developed firing technology, such as updraught kilns. In doing so, they have overlooked other ephemeral means of firing, which Feinman and Balansky found did exist, especially on sites that exhibited ample ceramic production but no formal kilns (Feinman 1997, 141).

KILN TECHNOLOGY DEVELOPMENT TABLE

PHASE	TECHNOLOGY	ADVANTAGE	DISADVANTAGE
1	Bonfire		
2	Shallow pit	Heat retention Higher temperature More control of temperature Slower cooling	Thermal stress Uneven heating Breakages
3	Low mud walls around pit	Step toward fixed structure Confined fire Slowed cooling	
4	Fire beneath wares	Circulation of heat More even temperature More efficient fuel use Step toward enclosed structure	
5	Covering the wares	Contains heat Decreases heat loss	Uneven heating
6	Separate fire chamber	Wares not subject to direct flame More controlled heating and circulation Fixed structure Can be used repeatedly	
7	Roof over kiln		

Development Sequence of kiln technology according to Rhodes.

Past archaeological researchers often espoused an evolutionary theory of kiln technology development, which followed a developmental sequence such as that outlined by Daniel Rhodes. Rhodes believed that "the kiln was not an invention but rather the result of a series of small improvements" (Rhodes 1981, 13). Such a sequence has been adopted by many art historians, archaeologists, and ceramicists, such as Glenn Nelson in his book *Ceramics: A Potter's Handbook* (Nelson 1971, 271-2). This early theory has been supported by many ceramicists, ceramic writers, and archaeologists, who believed that as knowledge and sophistication grew within primitive societies, basic pit kilns evolved into simple updraught kilns. The firebox was the sunken hole in the ground, upon which a firing chamber was built containing wares that were probably covered in broken shards and mud, leaving a small hole at the top for the smoke to escape. In the Far East, for example, simple updraught kilns evolved, being constructed into the sides of hills (Cooper 2000, 16). Such kilns, which theoretically evolved from firing in pits, allowed for greater control over draughts, atmosphere, and temperature, leading to the ongoing progress in kiln technology, and subsequent glaze and decorative technique development that we witness in contemporary ceramic practice.

Lindsay Scott, however, in his article "Pottery," suggests that the existence of rudimentary kiln technology did not necessarily imply that open firing or pit firing was no longer used. Rather, the size limitations of the constructed kiln, the larger quantities of fuel required, the fuel being more expensive (wood, rather than brush and cow dung, etc.), and the capital expended to build permanent kilns would have been prohibitive, except in major towns. Therefore, Scott argues, economic as well as traditional motives would have existed to continue firing in pits or open firing (Scott 1954, 392). Most modern research supports Scott's claim that the use of various kiln types occurred contemporaneously in prehistoric societies. Rice believes this has previously been overlooked because of modern society's preoccupation with technical progress and efficiency, both of which act as blinkers that "obscures our vision, rather than clarifies it" (Rice 1997, 252). This blinkering also increases and further complicates the limitations of archaeological methodologies discussed earlier. Heather Miller, in her discussion of the Indus, points out that there was variety in the types of firing structures used, and that while increasingly complex kilns were developed, variants of the older, more simple, structures continued to be utilized (Miller, 1997, 65).

Similarly, Izumi Shimada notes the excavation of a pottery workshop in Moche on the Peruvian coast. During the excavation, a circular pit kiln was unearthed as well as a rectangular above-ground kiln (Shimada 1997, 117). Why two kiln types, and how did these ancient potters choose one firing technology over another? It seemed that many early communities chose their firing method based on what and why they were firing rather than a developmental sequence of technology. As mentioned, within Indus communities, for example, newer technologies were developed, but older methods continued to be employed (Miller 1997, 65). Shimada was also involved in the excavation of a small ceramic workshop in Galindo on the Peruvian coast, which revealed a circular pit kiln of approximately twenty-one feet in diameter containing wood ash, llama dung, and large fragments of coarse utilitarian vessels (Shimada 1997, 116). The coarseness of the clay suggests the artisans chose that particular clay because it was to be fired in a pit firing, which can have uneven bursts of flame and heat, and coarse clay is more resistant to such thermal shock. In examining the choice of firing method within these ancient societies, Feinman and Balkansky refer to J. J. Sheehy's review of fifty-five modern ceramic communities, comparing resources, social organization, technology, and production aspects to formulate a model of ceramic production intensity. Sheehy noted a proportionate increase between firing frequency and the firing technology employed (Feinman, 1997, 141), so that for a society that produced a lot of ceramics or a large quantity of a particular item, it may have been worthwhile to construct and maintain more permanent firing structures, such as updraught kilns. In his examination of prehispanic kilns, Christopher Poole concludes that the rate of heating was a very desirable factor in firing choice. Speaking to a modern potter from San Isidro, Poole asked why she used a kiln instead of firing in the open as do the potters of Sehualaca. She replied that she needed to control the rate of heat to the pots and that she could not do this in an open fire (Poole 1997, 166). Poole then cites Gosselain whose research examined shards from the Matacapan Classic Period and found that less control was exercised over the Coarse Brown pottery than other untempered fine wares, indicating that the coarse ware was fired in pits, and the fine wares were fired in kilns (Poole 1997, 167). Poole concludes that the appearance of kiln technology seems to coincide with growth in popularity of fine paste wares and the uniform surfaces that can be achieved by a controlled atmosphere (Poole 1997,167). So, while an evolutionary kiln development theory is largely refuted today, and various firing methods coexisted, there is a noticeable trend toward more advanced kiln technology as production increased, finer wares became popular, and higher temperatures were desired. This indicates that the reasons for kiln evolution and innovation are more likely to be linked to social mores than societies becoming more technologically advanced..

CERAMIC TECHNOLOGY INNOVATION: HUNTER-GATHERER VS. AGRICULTURE

Prior to the discovery that clay hardened in fire, it had been long used for body painting, cave painting, strengthening building structures, and modeling ritual figurines used in fertility ceremonies, transition, and afterlife ceremonies. In fact, amongst the innovations used to mark the beginning of the Neolithic was the large scale use of clay within society, particularly in building (Moore 1995, 45). The Neolithic Period is generally considered to be the last phase of the Stone Age, marked by the domestication of animals, the development of agriculture, and the manufacture of pottery and textiles. The period is commonly thought to have begun around 9000 BC, ending with the Bronze Age, circa 3000 BC (Dictionary.com, accessed 24 July 2011).

It is believed the earliest ceramic firings occurred in and evolved from the tribal camp fire, and that once tribes adopted a less nomadic lifestyle, it is likely that firings initially continued in the same way, that is, firing of wares in a hole in the ground. It is possible, though, that some firing occurred in ovens found in and near village houses. In fact, it has been suggested that it was in pre-pottery Neolithic levels of stratigraphy at both Jericho and Jarmo, that humans may have discovered clay's ability to transform into ceramics. Here there have been found clay floors molded in one piece containing depressions that may have been used for basins, bins, or ovens. Ruth Amiran hypothesized that due to the closeness of these depressions and the tasks they represent—cooking, preparing, cleaning —that some sort of cross firing may have occurred, leading to the accidental firing of clay (Amiran 1965, 242).

Although the point at which it was discovered that clay hardened when exposed to fire is unknown, archaeological and anthropological knowledge of prehistoric humans enables speculation as to how this occurred. Steinen, for example, suggests that Native North American Indians kept clay for body painting in baskets. This clay may have hardened in the sun, providing a prototype for pottery bowls (Linné 1965, 36). Emmanuel Cooper also offers three distinct possibilities: fire being maintained in a hole in the ground; the role of myth and ritual in prehistory societies; and the use of bird's nests or baskets as utilitarian items (Cooper 2000, 9-10).

Fire was a precious resource to prehistoric humans, and offered a myriad of uses, the most important being warmth and light, protection from predators, and cooking. It is likely this resource was protected, and coals were probably carefully maintained by individuals charged with this responsibility. It is also likely that the fire was contained within a hole in the ground, where it was protected from the wind and rain, and covered to maintain coals. Such holes may have been waterproofed with clay, and over time, nomadic people discovered that clay hardened. When an agri-based lifestyle was adopted, this discovery may have been utilized; the hardened clay lining being used as a rudimentary vessel for cooking and storing food, and also as objects to accompany the dead into the afterlife.

Myth and magic played a major role in prehistoric nomadic humans' lives, and religious figurines were fashioned from clay for fertility ceremonies and rituals. Fire, being a powerful force, may also have been used in these ceremonies, so that figurines may have been placed in the fire, thus hardening the figure. Again, this knowledge would have been exploited only once humans adopted a less nomadic lifestyle.

Woven baskets, or perhaps bird nests, may have been smeared with clay to waterproof them, and foodstuffs placed upon them to cook on the fire. The clay would have baked and hardened. This method of cooking was possibly used for thousands of years, but the resultant hardened dish was only used and reused on a daily basis, as stipulated, once humans became less nomadic. Indeed, the application of this knowledge was crucial for the success of early agricultural tribes, as vessels and storage containers were essential items.

Cooper's three theories suggest that the basic knowledge that clay hardened in fire developed during the nomadic period, but was not fully utilized until societies became more agricultural, because it was of no practical benefit, as pottery was fragile and cumbersome, so not easily handled by a nomadic group. Recent research suggests, however, that some types of pottery was made and fired by nomadic groups.

In analyzing the development of pottery technologies, C.P. Garraty, in her article "The origins of pottery as a practical domestic technology: Evidence from the middle Queen Creek area, Arizona," defines two models: a culinary-subsistence model of pottery innovation and a social model of pottery innovation (Garraty 2011, 2-3). Social models highlight pottery as tokens of prestige or as technologies adopted to enhance one's prestige, as knowledge of such technology may have been viewed as superior to the common usage of gourds or baskets. She cites Karen Vitelli in explaining that the earliest ceramic artifacts were frequently imbued with sacred significance, and this may have been enhanced via the transformative power of fire. Based on these assertions, early pottery may have been used to increase authority and impress neighboring tribes or families during feasts or communal events. If this is the case, it would suggest that fired pottery was used by nomadic groups and that firing technology was more advanced than previously believed.

The culinary subsistence model emphasizes the development of ceramic containers as being tied to tribe mobility and a mixed foraging-agricultural community. Previously, it was believed that ceramics were too heavy and burdensome for nomadic tribes to carry; however, tribe mobility was cyclic and based on seasons, so that ceramic containers may have been left at the site, ready for the next season. In such communities, containers may have been used for both storage and cooking. The trend toward a mixed economy also may mean that women were the main innovator in pottery development due to an increase in family tasks, such as gathering wild plant foods, tending to gardens, food preparation, child-rearing, laundering, and other labor. Women may have responded to this by weaning children earlier, and pottery containers were much more practical than a fire pit or organic containers, such as gourds, to cook down grains and seeds suitable for weaning (Garraty 2011, 4). Again, this theory implies knowledge of ceramic firing that was previously not considered, probably because, as noted earlier, there is a lack of surface indicators associated with firings and pit firings that are used infrequently and that have no permanent structure.

Garraty's culinary subsistence model suggests that women were in the prime position to discover and develop the use of ceramic wares within their communities, and many of the shards found on archaeological sites indicate vessel shapes were dictated by domestic use. According to researchers such as James Mellaart and Lindsay Scott, these domestic forms were already in existence, being informed by the pre-pottery forms of wood and stone vessels and basketry (Mellaart 1965, 220; Scott 1954, 397). Why then did pottery supersede the use of these materials for domestic vessels? Mellaart hypothesizes that in places such as Catal Hüyük—a very large and well preserved Neolithic settlement in southern Anatolia, which existed from approximately 7500 BC to 5700 BC in the center of an alluvial plain, with no supply of workable stone for about thirty miles [fifty kilometers]—would have stimulated the beginnings of pottery production.

In Khirokitia, Cyprus, however, early attempts at pottery making were abandoned for the use of stone (Mellaart, 1965, 225). This may have been due to several factors, including aesthetic, religious, and firing difficulties, and illustrates the influence of environmental factors in the invention and advancement of pottery and firing techniques.

As indicated, before the discovery of pottery-making, human hands had already developed the dexterity and skill to carve and form vessels in wood, stone, and basketry. This process meant that functional shapes had been invented. Humans already possessed the knowledge to mix clay and model figurines and ritual objects. Once they discovered that clay hardened, therefore, all that was needed was the transference of pre-pottery knowledge to clay, and the continued exploration of firing techniques. Much of this experimentation may have occurred in the location where much of the firing already took place: the cooking oven, and therefore may have been mostly carried out and developed by women.

EXCAVATED PITS

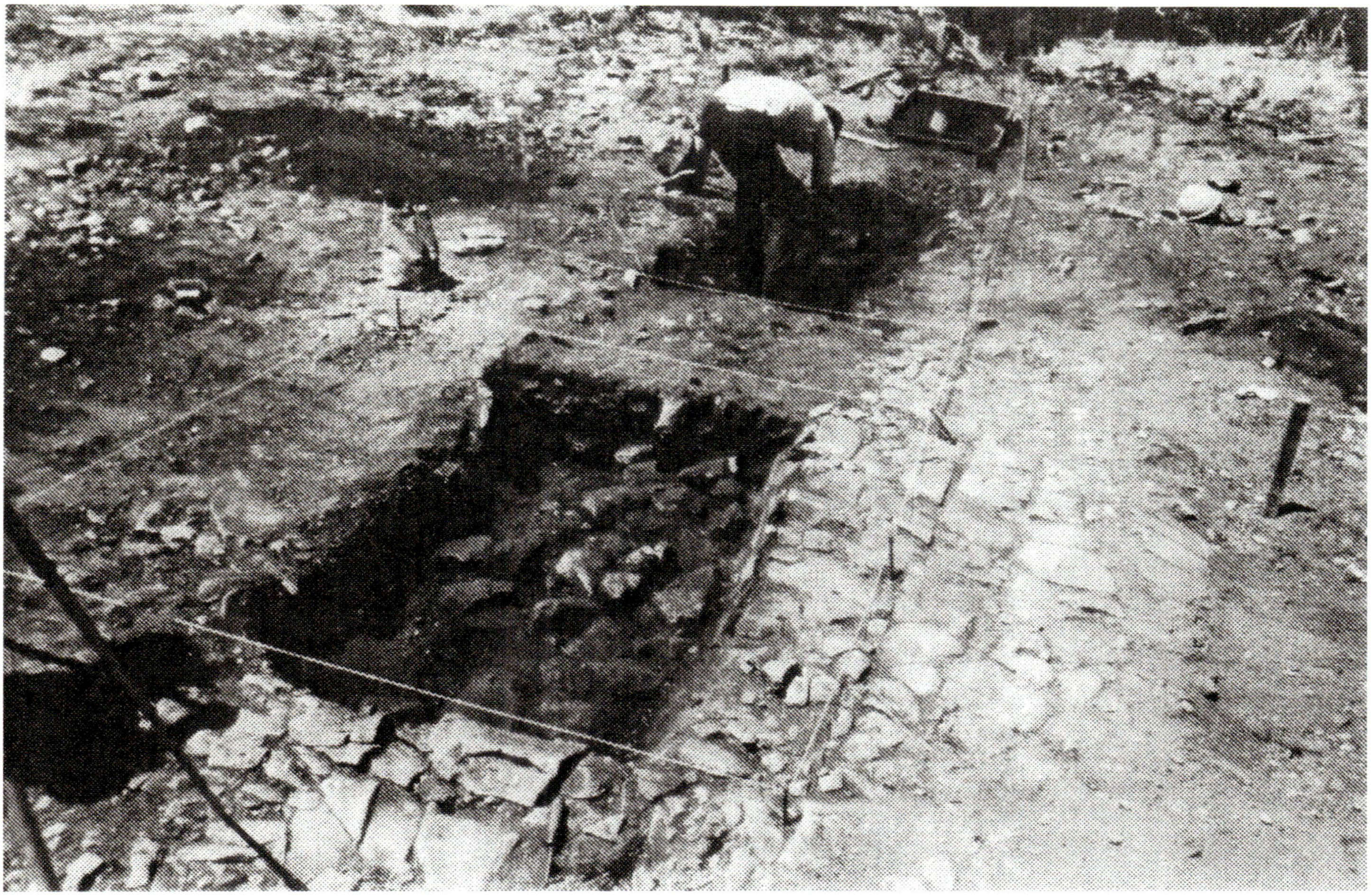

Trench Kiln under excavation by La Plata Archaeological Consultants in 1982. *Reprinted with permission of The American Ceramic Society, www.ceramics.org. All rights reserved.*

Despite the obstacles confronted by archaeologists in recognizing pit kilns, discussed earlier, many pits have been excavated from archaeological sites, especially in more recent times, as archaeologists have become increasingly aware of the inefficiencies of earlier methods of discovery and classification. More recently, archaeologists have employed the expertise of potters in order to recognize certain elements in a survey site, and to help duplicate the building and firing of prehistoric excavated kilns, in an attempt to understand how they may have operated within the production process.

During the early 1980s, archaeological excavations uncovered several "trench" kilns in the Mesa Verde and the Northern San Juan region in South America, which were dated between AD 900 and AD 1300. These kilns were completely different in design and scale than any firing structures that had been found previously, so new models of interpretation had to be developed. In 1991, the Crow Canyon Archaeological Center in Cortez, Colorado, hosted a conference that included archaeologists, Native Americans, and potters. The intention was to study and reinterpret ancient Anasazi firing techniques (Blinman 1997, 86). These kilns were generally stone lined and rectangular in shape, and contained remnants of charcoal, partially burned fuel, blackened stones, and shards (Ibid, 88). As more kilns were identified, Mesa Verde National Park archaeologists wanted to develop models for how these kilns functioned, and found they made the most progress when they collaborated with artist and potter, Clint Swink (Ibid, 90). In order to reproduce a degree of authenticity, firing duplications were based on trench design, raw materials, and fuel remnants found in the trench kilns, and the appearance of Anasazi pottery. While researchers admit their experimental firing duplications may not be exact, they do provide insight into the firing regime of the time.

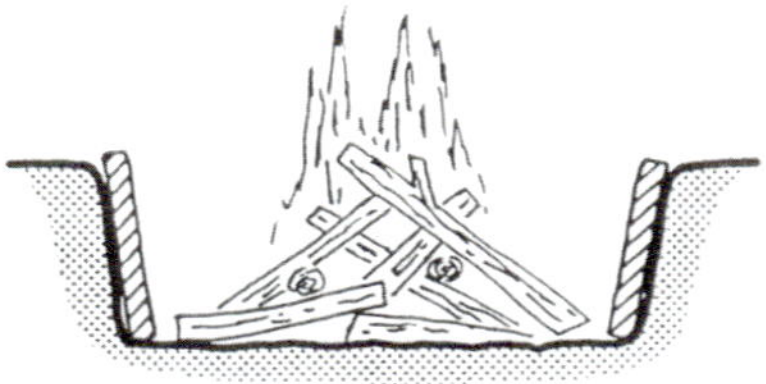

Preliminary fire – A fire is constructed within the kiln and allowed to burn to produce a layer of charcoal across the trench base.

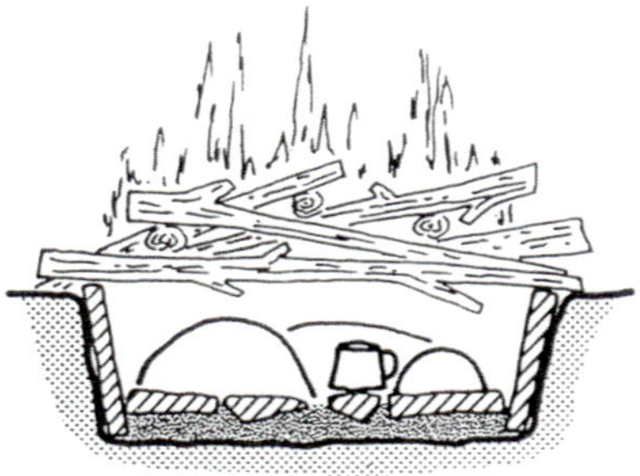

Secondary fire – The stone platform, vessels, and cover shards are placed on the charcoal layer. Fuel is ignited over the setting.

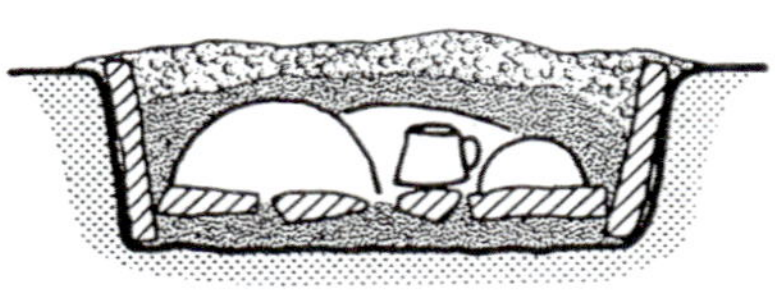

Quenching – After the fuel burns to charcoal and collapses, the setting is covered with a smothering layer of earth and allowed to cool.

Firing sequence model based on trench kiln replication experiments. *Reprinted with permission of The American Ceramic Society, www.ceramics.org. All rights reserved.*

Archaeologists have also sought potters' advice on why certain elements were used in firings or why certain clays were used. Remains of a primitive local plant with segmented, hollow stems and corn-like spikes called Horsetail were found in two circular pit kilns unearthed at Maymi in the lower Pisco Valley by Martha Anders and her colleagues. According to Peruvian potter, V. Chang, traditional potters on the south coast of Peru place a layer of dry cow dung mixed with horsetail above and below their pots to be fired, because the silica in the horsetail aerates the dung, causing it to burn evenly and thoroughly (Shimada 1997, 116).

During the 1990s, pit kilns were found at Mochica and Nasca (ca. AD 450–750) on the Peruvian coast. One pit, situated on the sloping side of a gully in order to take advantage of wind for draft, measured approximately 8' 2" in length and 10" in depth. Charcoal remnants in the pit indicated fuel type and discoloration of the pit illustrated high temperatures, as did over-fired shards found at the site. This ability to reach a high firing temperature was confirmed when compared to a modern pit of approximately the same size and shape, and using the same fuel, in Mórrope, 106 miles from the site, which recorded temperatures over 1920°F (Ibid, 115).

A 1991 surface survey in Pampa de los Burros revealed a ceramic workshop consisting of a large circular pit kiln of approximately 29' [9 m] in diameter and eighteen small masonry rooms. The pit was 2–4" deep and filled with dark ash, charcoal, and shards. Within the pit was a smaller, intensely reddened pit of 21' x 15' x 15 ¾" deep, which was full of compacted animal dung (Ibid, 120).

Pit kilns were also excavated at the Inka ceramic workshop at La Viña. The workshop contained several circular, shallow, reddened pits filled with ash and charcoal remnants (Ibid, 122).

In the early 1990s, a pit kiln was found at Harappa dating to 2300 BC. In writing about this kiln Heather Miller is curious that more pit kilns have not been found in the area, given that pit firing is currently the most common form of firing practice in South Asia. She points to pits previously not being recognized or mistaken for cooking structures as the reason for this oversight (Miller, 1997, 51). The Harappa pit was 31 ½" x 27 ½" x 12" deep, and its teardrop shape indicates an opening for air and fuel. The pit's base was covered with a thick layer of ash and a layer of pot shards. Based on local contemporary potter's techniques, it was decided the shards were used to protect the wares from direct flame during firing. This pit also contained low-fired triangular terracotta cakes in the pit opening, and again, based on modern Pakistani potters' techniques, it was deduced that the cakes were placed at the mouth of the firing structure so as to control air flow, reduce fuel consumption, and maintain an even heat (Ibid, 51-2).

During the early 1990s, excavations at Ejutla uncovered the stone foundations of a house along with evidence of ceramic and other craft production (Feinman 1997, 134). This evidence included large amounts of shards and production related ceramic products, but no formal kiln structures. This led researchers to suggest that firing at Ejutla was performed in pits. After further excavation, several bedrock pits were found filled with dense midden deposits and shards (Ibid, 136). The best examples were asymmetrical, ranging in size from 6 ½' to 13' across and 15 ¾" to 27 ½" deep. The irregular shape suggests a mouth for stoking and air flow control, and each was orientated south, the direction of the afternoon wind, to take advantage of draft (Ibid 1997, 139).

The continuation of pit kiln technology can still be observed today in many places throughout the world. Writing in 1965, Hélène Balfet noted that amongst the various groups of pottery-makers in Maghreb, especially "nonspecialist" women, many still employed methods to reduce the difficulties of kiln firing. This included utilizing, "a flat surface (sometimes limited by a ring of stones), a slight slope, or a true pit holds the pottery, which is very carefully placed on a bed of fuel. Above and around it is added the rest of the fuel—wood, straw, or other dry plants, or dung cakes" (Balfet 1965, 166-7).

CHAPTER TWO
CONTEMPORARY HISTORY

> Whether motivated by energy savings or not, ceramicists have recently moved in directions which require less heat ... Another is the enthusiasm for open pit firing, using wood or other fuels of little value. It is widely accepted now that neither beauty nor utility in the finished object is dependent on high firing (Rhodes 1981, viii).

CONTEMPORARY PIT-FIRING

During Owen Rye's field trip to Pakistan in 1971–72, he came across a book entitled *The Potter's Book*, which connects the daily life of the potter to the practice of Islam. Each ceramic task, from digging the clay to unloading the kiln, is associated with a prayer to Allah, and while the origin of this book is unknown, according to Rye, it is thought to date back to Persian literature (Rye 1976, 191). It is through this book that we have a hint that early firing practices in Pakistan were carried out in a pit.

> When you put them in the kiln *and heap fuel on them* you should recite these prayers and leave the rest to Allah: "O Allah save them from all calamities and *damage and from the rain*" (my emphasis) (Rye 1976, 192).

While in Pakistan, Rye observed the contemporary firing practices of local Pakistani potters, practices that have been handed down over generations to the point where, when Rye asked one potter if he used

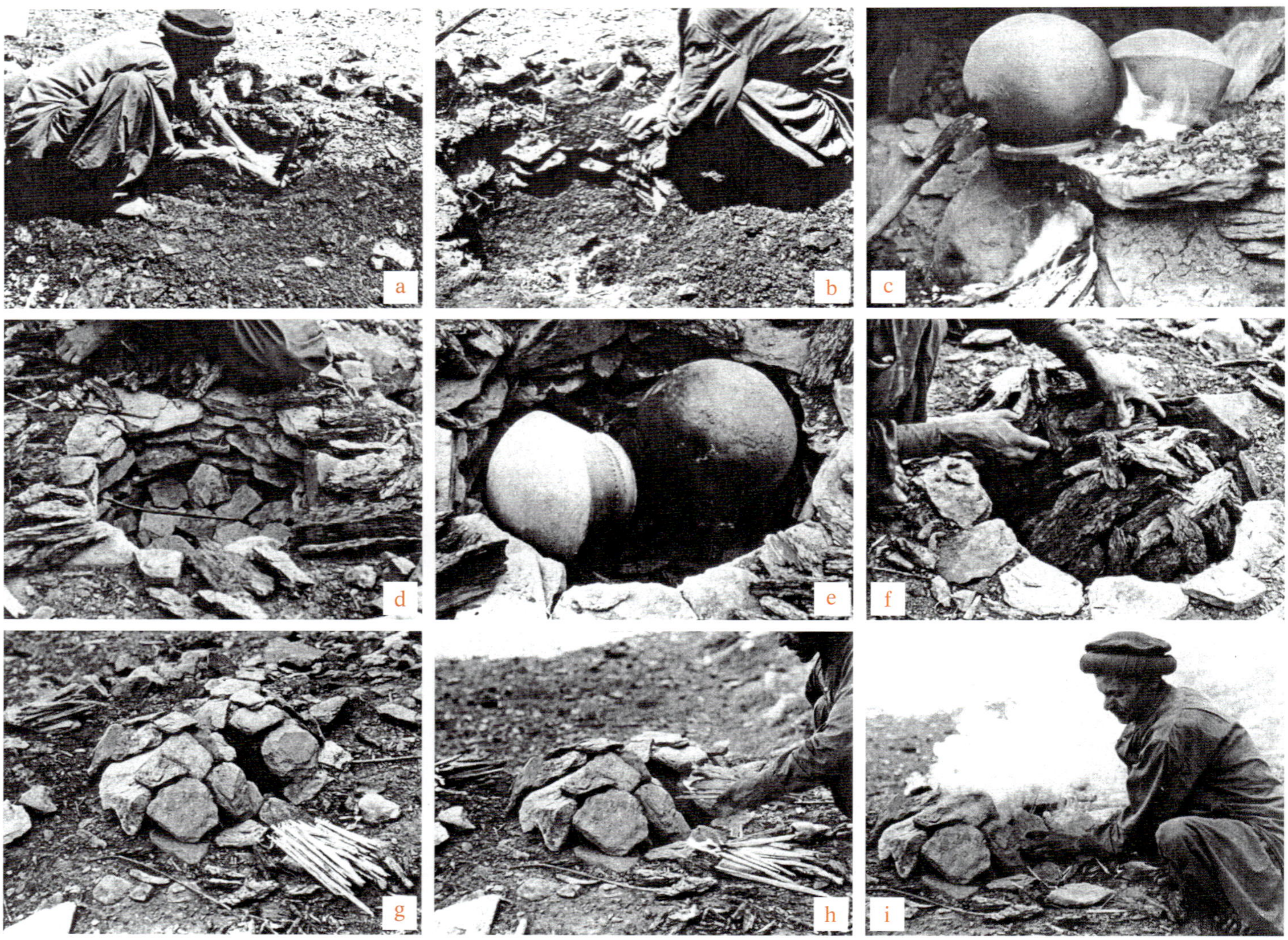

Bomboret potter's pit firing method: a. excavating hole for the pit kiln, b. lining the hole with pieces of slate, c. preheating vessels over a fireplace in blacksmith's shop, d. slate-lined pit, e. vessels placed on bark fuel in pit, f. more bark placed around vessels, g. vessels and fuel covered with slate, h. igniting the fuel with kindling, i. sealing the pit completely with slate after fuel has bugun to burn. From: *Traditional Pottery Techniques of Pakistan*. *Photos: Owen Rye*

a certain fuel because of the kiln design, or if the kiln had been designed for that specific fuel, the potter did not know—he merely fired in this way because it had always been this way (Rye 1976, 166).

Rye has broken down the firing practices in Pakistan into two categories: unglazed ware and glazed ware. Unglazed ware is generally fired in what Rye calls a "mixed firing," where there is no separation of the fuel from the wares, and within this category many Pakistani still use pit kilns.

A potter Rye visited in Bomboret in Northern Pakistan uses a pit that he digs anew for each firing depending on the number of pots he has to fire. One can imagine how such a method would leave little trace for future archaeologists. The pit is lined with slate, and after preheating the pit and pots separately to minimize thermal shock, the pots are placed inside the pit and gaps between pots are filled with pine bark. The pit is then covered with slate for insulation (Ibid, 165).

Rye's travels also led him to areas along the Pakistan-India border, where he noted the *ãvĩ,* a type of kiln he found highly prevalent in Panjab and areas west of the Indus. The *ãvĩ* is an open, shallow pit, triangular in design and built on a slope. Fuel and pots are loaded in layers and the fire burns gradually upwards through the wares. Fuel varies depending on what is available, though dung is favored as it burns the most evenly (Ibid, 166).

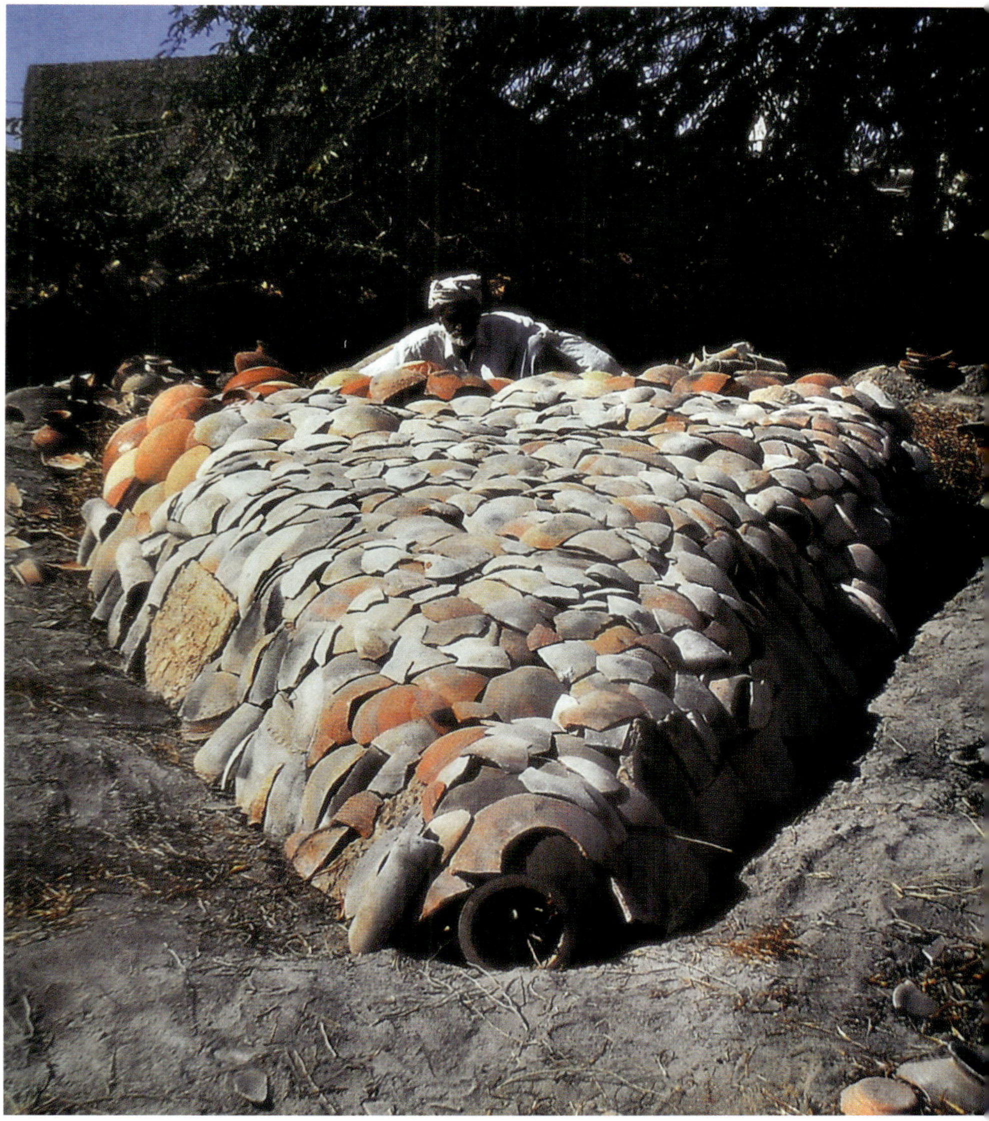

Avi style pit kiln: the pile of three hundred pots is covered with sherds. This will be followed by a thick layer of straw and then a layer of ash to keep the straw in place and conserve heat. Buffalo dung has been packed into the apex of the triangular pile and will be ignited through the neck of the water pot sherd with straw. From *Traditional firing in India. Photo: Jane Perryman.*

Jane Perryman has also traveled extensively through India, researching the traditional pottery techniques undertaken by contemporary communities, including firing methods. While many potting families fire in open areas, some do use pits in which to place their wares, and she describes a pit kiln in Kutch similar to the *ãvĩ* Rye found prevalent west of the Indus during his fieldwork twenty-five years earlier. Lodai is a small town in Kutch that at one time boasted eleven families manufacturing pottery. Due to declining demand, this has decreased to five families that provide the town with cooking, storage, water, and ceremonial vessels. Some vessels are also traded and taken to fairs. The making cycle involves the entire family: women collect and mix the clay, men throw the pots, slip decoration is applied by the women, and the firings are undertaken by the men. The particular family Perryman studied carry out eight firings per year. The kiln site is raised up a couple of feet on a bed of ash, and is an elongated triangular shape, twenty feet long by twelve feet wide at the back moving into a point at the front. While the pit is being prepared, the awaiting pots are warmed, with children systematically rotating them to face the sun. A fourteen-inch deep pit is dug from the bed of ash, with the ash being piled around the edges of the pit to form a twenty-four-inch high wall. A foot thick layer of sesame seed straw is compressed into the bottom of the pit and a layer of *medival* straw, from the *medival* plant, an ayurvedic laxative locally cultivated, is spread on top. Next, a layer of thorny brushwood, indigenous to the area, is added, and all layers are then compressed into the pit by stamping up and down on the pile. More brushwood is then added and compressed with a rake. The pit is then ready to load and the children carry the pots to the pit according to their size and type. The kiln is loaded from the back, placing pots upside down on a shard. The kiln takes a few hours to load, as the pile of pots is thirty-five inches high and contains three hundred pots. More straw is added to the pile and the entire stack is covered in shards. A trench is dug all around the triangular shape, the ash being thrown over the shards. A space is left at the front of the stack and packed with dry buffalo dung. In front of this, a broken neck of a water pot is placed, facing outwards. Some straw is lit and placed in the mouth of the shard. The firing burns slowly for approximately fourteen hours, and then takes twenty-four hours to cool (Perryman 2000, 82-5).

Vara is a small town about fifty-five miles from Bombay, and the pottery making season is from October to May, during which time vessels are made for festivals and ceremonies, storage and water. Like most parts of India, and other traditional pottery manufacturing communities, the industry is shared by the extended family. Here, however, the actual making is undertaken by the women of the family. Perryman speculates that this may be why there is no surface decoration or embellishment on these wares, as there is insufficient time. Finished pots are burnished with beans from the *arita* tree, which are threaded on strands resembling a necklace and dipped into kerosene (Ibid, 47). The firing pit is shared by three to four households of the same extended family, and is situated on the outskirts of the community. It is rectangular, and approximately sixteen feet long by seven feet wide and sloping to three-and-a-half feet deep at the back. The pots are laid out in the afternoon sun to warm. While the pot making is carried out by the women in this community, the men take charge of the firing. The pit floor is covered in rice husks, which are formed into furrows along the length of the pit with twelve-inch gaps in between to form stoking channels. Bricks are placed in the channels to support the larger pots and allow heat to circulate under them. Smoldering dung cakes are placed at two-foot intervals along the furrows. The largest urns are placed on the bricks, neck upwards and angled away from the front of the pit. A stack of fired water pots is then placed in line with the large pots, leaving gaps to create stoking channels. The rest of the pots are then stacked inverted on top of the larger pots. The stack is finally covered in a layer of rice straw, which is compacted with a bamboo pole. The straw is prevented from flying away during the firing with a mixture of ash and rice husks thrown on top. Sticks are placed to cover the stoke holes, and the front is covered with rice straw also. The loaded pit is then left to preheat for four-and-a-half hours, as the smoldering dung will slowly ignite the rice husks. The firing is performed during the night so that the temperature can be monitored by its red glow. Gradually, small pieces of wood are pushed through the stoke holes, then eventually longer pieces until wood is pushed through to the back wall of the pit. After about two hours, the color at the front is satisfactory, but the back is still not to temperature, so wood is stoked through the straw till the correct glow color is reached. The pit is unloaded the following day, and losses are minimal (Ibid, 50-1).

The small town of Nizamabad, in the densely populated Uttar Pradesh, was invaded and colonized by Muslims in the seventeenth century, and the forms and surfaces of many of the pots here are noticeably Muslim influenced. There are two levels of production—domestic wares, including pipes, incense burners, oil lamps, and pots, and decorative black-fired items, which are exported. Though not many family orientated traditional potters come to recognition in India, due to caste prohibitions, Ragendra Prajapati's black-fired wares are National Award winners, and his work is sold through the government emporia and is exported abroad. Ragendra's works are wheel-thrown, incised and burnished with an agate pebble, then after firing are inlaid with a mixture of lead, zinc, and mercury. The firing pits are housed in a thatched hut where the floor is covered with a thick layer of ash from previous firings. An eighteen-inch deep pit is dug in the ash and lined with buffalo dung, which has been aged for a year to maximize heat generation. The firing chamber, a large pot originally used to store grain, is fitted inside the pit. Some dry, crumbled dung is added to the chamber and works are gradually packed, adding dung in and around the wares as the packing continues. Some crumbled dung is lit from the previous night's firing, which is added to the chamber. A lid is fitted and sealed with a mixture of ash, dung, and water. The portion of the chamber still protruding from the pit is then covered in dung. A brick line is placed around the circumference of the pit, leaving a gap for stoking, and shards are leaned and stacked on the bricks about two-thirds up the chamber. Ignited dung is placed beneath the kiln, and the opening covered with shards. The entire chamber is finally covered with damp ash, which is compressed against the surface to help hold the fuel in place and retain heat. Two hours later more damp ash is compacted around the chamber. After two-and-a-quarter hours, some shards are removed and tongs are used to remove the lid, dried goat dung is added, and the lid quickly replaced and resealed. This process is repeated again after five minutes, causing a reduced and carbonizing atmosphere within the kiln. Finally the chamber is covered in compacted damp ash and left to smolder and smoke till the following morning. The results are highly glossy black pots, and breakages are rare. Inlay work is carried out post firing (Ibid, 66-72).

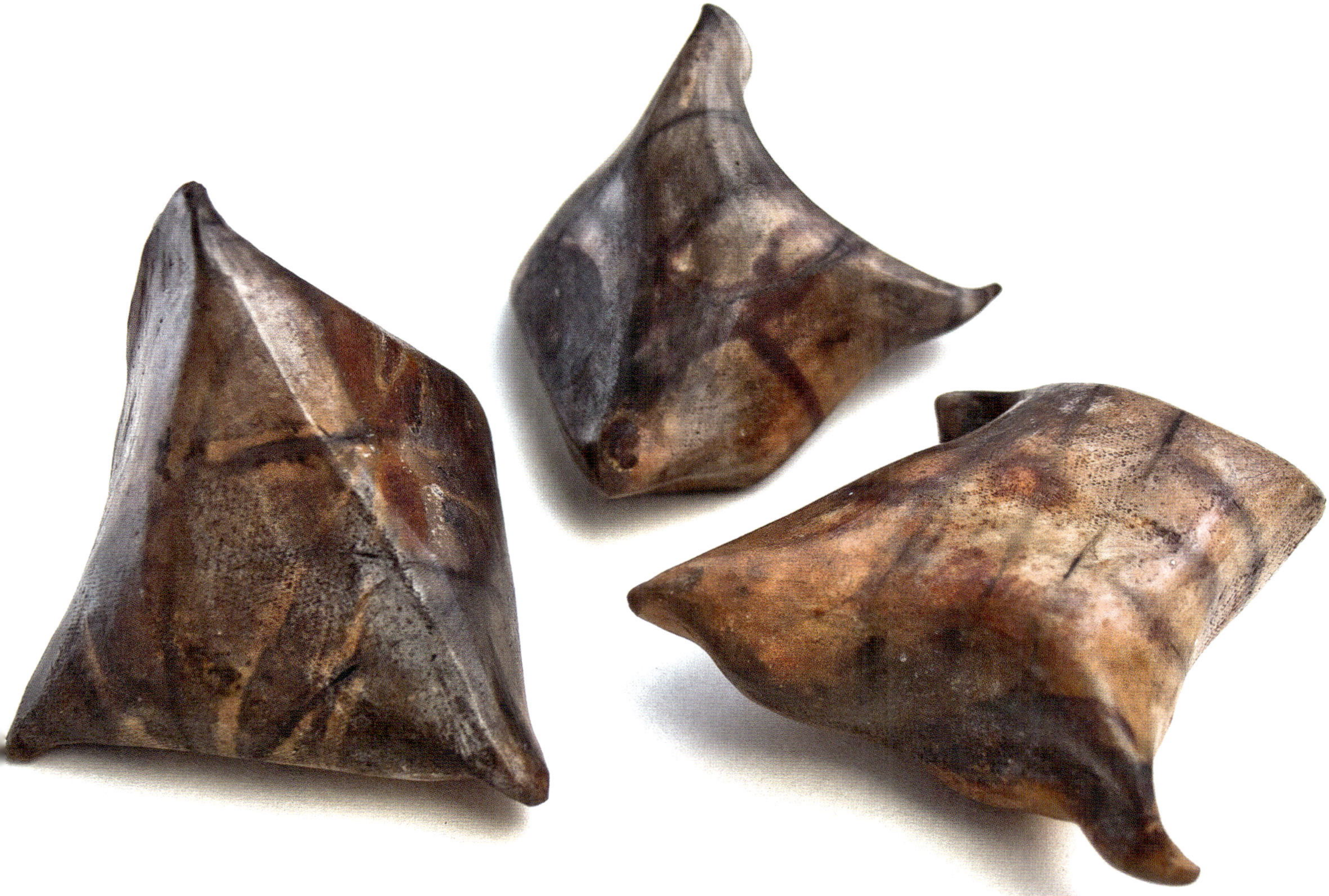

Shazieh Gorji. *Listen, stones have stories too,* 5" x 5". Hand built, burnished, wrapped in copper wire, banana peels, and salt soaked twine. Surrounded by copper carbonate, red iron oxide, and turmeric. Pit was fed wood for more than half a day, after which it was left to burn off and cool overnight. *Photo: Saleema Hamid*

In her book *Approaches to Archaeological Ceramics,* Carla Sinopoli examines a traditional pottery making family in a large village in southern India. As with the previous examples, pottery making is traditionally inherited from relatives, and is practiced by the entire extended family. Wares are fired in an open pit surrounded on one side by a permanent semi-circular wall. Stacking and ventilation is precise. Fuel is layered on the floor and the sun-warmed pots are stacked with shards separating and supporting the pots. Once all the pots are stacked, circular vents are made from broken vessel necks. A final layer of shards, which is covered with more fuel, is added carefully, without covering the vents. The entire pit is then covered with a layer of ashy soil. The fuel is lit from the bottom using a stoke hole in the stone wall. The firing lasts approximately three hours, and then all the vents are closed. This method causes a reduction atmosphere in the pit and produces darker, carbonized pots. If red pots are preferred, the vents are allowed to remain open so that an oxidized atmosphere is achieved (Sinopoli 1991, 38-9).

Traditional pottery making and firing techniques are disappearing due to globalization, the cost effects of mass production, the availability of plastics, stainless steel, and refrigeration, and new limitations that are being implemented in places such as India, where the government has introduced restrictions on the amount of clay that can be taken from some clay collecting areas and lakes due to ecological reasons (Perryman 2000, 67). Ironically, in places such as India and Pakistan, lack of access to some modern technologies and resources contributes to the tendency of some contemporary ceramicists to experiment with traditional techniques.

Traditional Contemporary Potters

Shazieh Gorji is a Pakistani artist who has been working with clay for seven years. She graduated in 2004 from Bennington College in the US, where she studied Visual Arts with a focus in Ceramics. Her works consist of both functional pottery and sculptural forms. While in the US, Gorji assisted various wood fire ceramic artists and potters and found herself drawn to the glazing effects achieved by raw flames, days of firing, and melting ash. On her return to Pakistan, she built a raku kiln, and also began exploring traditional methods of firing because she lives in a country where it is difficult to source high-fire materials, kilns, and enough gas pressure to successfully reach temperatures beyond 2010°F. While visiting village potters and local craftspeople, she observed they all fired their earthenware pots in pit or mound kilns. Firing to low temperatures, they would often only bisque their functional works or use economically viable lead glazes. Understanding the need to continue her ceramic processes without compromising her concepts and glaze safety, Gorji built a pit kiln in her garden and began experimenting with techniques to achieve color on her sculptural works. This has proven to be a successful alternative and she is currently exploring ways to incorporate the traditional firing methods of the region with a contemporary twist. *Listen, stones have stories too* is a commentary on changing landscapes. "Influenced by my surroundings, travels, and observations, I try to listen to the stories told by stones, rocks, pebbles, shells, and sticks I come across. Enduring time and natural phenomenons, stones can almost be ignored as a common element of the landscape. Their stillness contrasts the varied textures, layers, and streaks of color present within them."

Despite the obstacles discussed above, such as globalization, local restrictions, and access to fuel, pit firing continues to be undertaken in communities that still use pottery as a connection point for family, community, ritual, and ceremony. These countries include India, Papua New Guinea, South America, North America, and Pakistan, though in some parts of North America, such as the Southwest, much of the pottery production is probably motivated more by tourist demand rather than community, as illustrated by the commercial trends seen in Pueblo pottery production. As a result, many Native North American potters have become collectable artists, such as Maria Martinez and Peter Jones.

Peter Jones was born into the Beaver Clan of the Onondaga Nation at the Cattaraugus Reservation in Seneca Territory (northwest New York State). He uses pit firing because it was the method used by his ancestors—at least that is the evaluation he has been given by archaeologists who have conducted research in this area. Jones's artwork is influenced by Iroquois pottery, which he became interested in exploring in 1977, upon his return from the Southwest, where he studied under Hopi artist Otellie Loloma while attending the Institute of American Indian Art in New Mexico. Iroquois pottery died out completely in the late 1500s, when it was replaced by copper and iron kettles that were more durable. The art of making clay pipes continued but was short-lived. The lack of information about Iroquois pottery methods led Jones to experiment with various making and firing methods. Using what he had learned from the Southwest potters and his own knowledge of pottery, he developed his current style of neo-traditional Iroquois pottery. Jones has exhibited widely and his work is held in several museum collections in the US.

The previous two artists illustrate a contemporary approach to a traditional firing technique being undertaken within a customary environment, and both artists are aware of the historical roots that stimulate and enlighten their research. Much of these traditional techniques discussed are used to fire unglazed wares for functional use, but in some areas, such as Quetta, Pabbi, and Hala in Pakistan (Rye 1976, 21), glazed ware is also produced, so that in these areas multiple kiln technologies exist. Since ceramic practices are traditional and hereditary in Pakistan, the existence of multiple kiln strategies gives credence to the contemporaneous evolution of kiln development theory explored earlier.

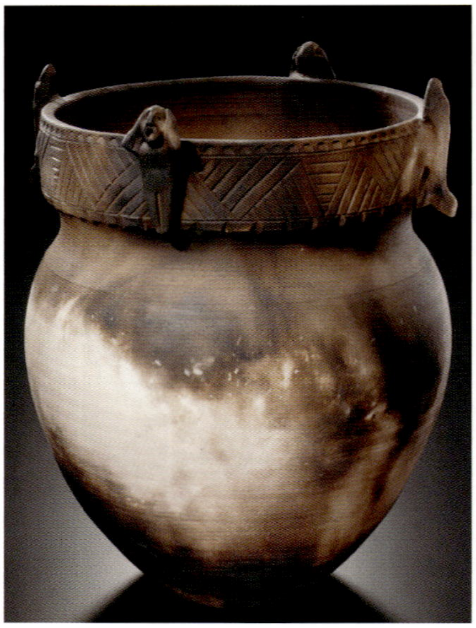

Peter Jones. *9-11,* 14" h x 9" w pot commemorating the twin tower attack. *Collection of the National Museum of the American Indian, Smithsonian, Washington D.C.*

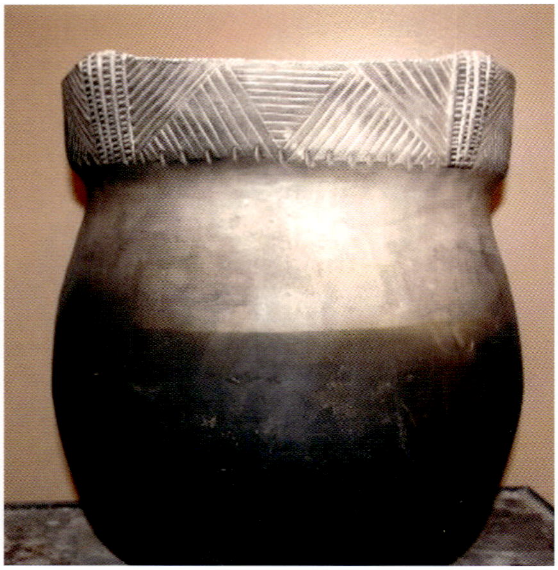

Peter Jones. *Seneca Cook Pot,* 20" h x 12". Private collection. *Photo: Peter Jones*

Reinterpreting Ancient Traditions

This multiple use of kiln technologies is prevalent in modern ceramic practice, with many ceramicists using multiple forms of firing depending on what they are producing. American ceramicist Robert Compton produces ceramics that are fired in a variety of methods. His functional ware, for example, is fired to stoneware temperatures; however, he incorporates a range of firing techniques into his art practice because he believes that using various methods keeps his enthusiasm as a potter alive. This approach also broadens his market appeal. Similarly, English ceramicist Marvin Kitshaw embraces several firing techniques in his studio practice.

The inclusion of pit firing into many contemporary ceramicists art practices, however, is a relatively recent event, dating from the 1960s and '70s. The development of a sophisticated kiln technology has resulted in the modern ceramicist becoming increasingly removed from the direct practice of firing. Modeling objects from clay is a hands on activity, as are decoration and glazing. Much ancient knowledge and many techniques have disappeared with the emergence of technologies that gave rise to plastic, stainless steel, and mass production. This same technology witnessed the advent of the electric kiln, pyrometers, and computer automation, all of which, it can be stated, lessen the excitement, spontaneity, and immediacy of firing from ceramic practice. Electric kilns are automated, simple to use, achieve reliable and predictable results, but as noted by ceramic commentators Sumi von Dassow and Jane Perryman, these technologies alienate the potter from the whole experience of making pottery, and they contend, as do I, this is a major reason for the increasing interest in alternative firing practices, such as pit-firing, that use fuel and flame, amongst contemporary ceramicists (von Dassow 2001, 1 & 44).

Within the context of the twenty-first century, the continued interest in pit firing may also be attributed to an increasing awareness of the issue of sustainability. Electricity and gas are nonrenewable resources that also release pollutants into the atmosphere. Pit firing, on the other hand, can utilize garden and bush property debris, kitchen scraps, cow dung, offcuts from timber yards—the list goes on. Trees can also be planted to offset the use of timber yard woods.

PIT FIRING IN AMERICA

While Bernard Leach did visit America with his friend Shoji Hamada in 1952 to spread his philosophy about pottery making, it did not take as firm a hold as it did in other parts of the world, such as England or Australia. This was probably due to the population of artists, craftspeople, and designers that had migrated to America during the 1930s and '40s, escaping the tyranny of a war-torn Europe. This creative deluge set the scene for a post-Dada avant-garde movement within the American art scene, and this included ceramics. This would have made potters and artists at the time less receptive to Leach's conservative philosophy. American potters did, however, find Leach's introduction of Japanese raku an exciting concept, and according to von Dassow, this led to the use of pit firing within American contemporary ceramic practice.

After World War Two (WWII), most Western economies prospered, and this meant increased income and leisure time. Many art courses were offered to both ex-service people and members of the public. Since most domestic/functional ceramic ware had been produced in factories since the industrial revolution, most experienced ceramicists found positions teaching at art schools, community centers, and universities. In America, this resulted in many ceramicists concentrating on producing one-off exhibition pieces, rather than the production ware of the studio potter, and to stand out in exhibitions, these pieces had to assume a different entity—they had to be larger, textured, and bolder (Nelson 1971, 135), an aesthetic that fit within the abstract expressionist movement that was also rising in America. Hence there was a departure from the pre-WWII functional thrown wares to hand built, thrown, textured, and non-functional expressive forms. Potters suddenly considered themselves artists and the distinction between "pottery," "ceramic sculpture," and "fine art" has been contentious ever since. This movement was to dovetail perfectly with the political and social upheaval that influenced the arts in the 1960s and '70s. During these decades, within the art world, questions began to emerge about the meaning and function of art. Artists began to experiment with different materials and methods of presenting their work. Some artists, such as Picasso, began to introduce clay into their oeuvre, some using the clay surface as a palette, and others using the plastic qualities of clay to express political and social commentary. The Funk Movement of the Bay Area led by Robert Arneson and the abstract expressionist use of ceramics initiated by Peter Voulkos popularized the application of clay as a material. Such breakthroughs laid open the pathway to exploring new methods of making and firing, and finding meaning within the ceramic world.

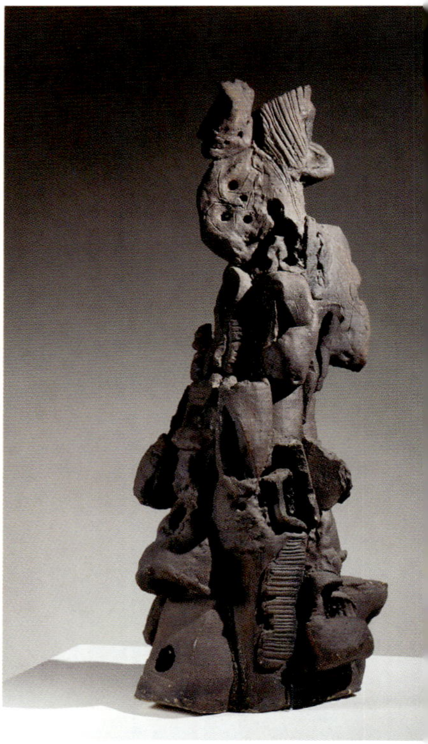
Peter Voulkos. *Tientos*, 1959, stoneware with slip, gas fired, 55" x 19" x 24". *Collection of the San Francisco Museum of Modern Art; purchase through a gift of Phyllis Wattis and gift of Gregory LaChapelle, 1998. Photo credit: © schopplein.com. Courtesy of the Voulkos & Co. Catalogue Project.*

The political and social turmoil of the 1960s and '70s resulted in people questioning industrial processes and the alienation many social commentators believed resulted from it, and indeed many of these questions still exist today as a response to the electronic age. According to von Dassow, the evolution of modern pit firing began in the 1960s as potters looked back to pottery's pre-industrial roots. Raku intrigued American potters because it offered a more immediate pottery experience, and ceramicists such as Hal Riegger began to experiment with pit firing as well as raku as a means of gaining insight into early pottery methods. Other raku and wood firers began leading workshops where participants used local clays and bisque fired them in large pits prior to glazing and refiring in brick or stone kilns built during the workshop. The smoked effects on these bisque pots prompted some participants to explore pit firing further (von Dassow 2001, 9). As noted earlier, Peter Voulkos was at the forefront of this experimental mood, and his extensive oeuvre includes pit-fired works. At the same time, the investigation of pottery's pre-industrial history prompted an interest in American Indian pottery techniques. American potters began to dig their own clay, incise and burnish and primitive fire their pots. A technique used by Pueblo Indian potters is to smother the fire to produce black pots. This practice may have prompted the use of post fire reduction in American raku. Thus, von Dassow argues, raku and Pueblo Indian pottery led to an experimental investigation into pit firing and the conclusion that that this phenomenon was a uniquely American one (von Dassow, 2001, 9).

Questions of industrialization and consumerism also influenced changes on the 1960s and '70s art scene. Artists began to question the role of art galleries, the art object, and the isolation of the studio artist. This questioning lead to new developments within the art arena, such as land art, installation, collaborations, and happenings. Within this context, pit firing became popular because it fit this inquisitive atmosphere: pit firing created non-functional objects, it occurred outside the traditional studio and gallery space, it was a communal and collaborative experience, and it was a happening.

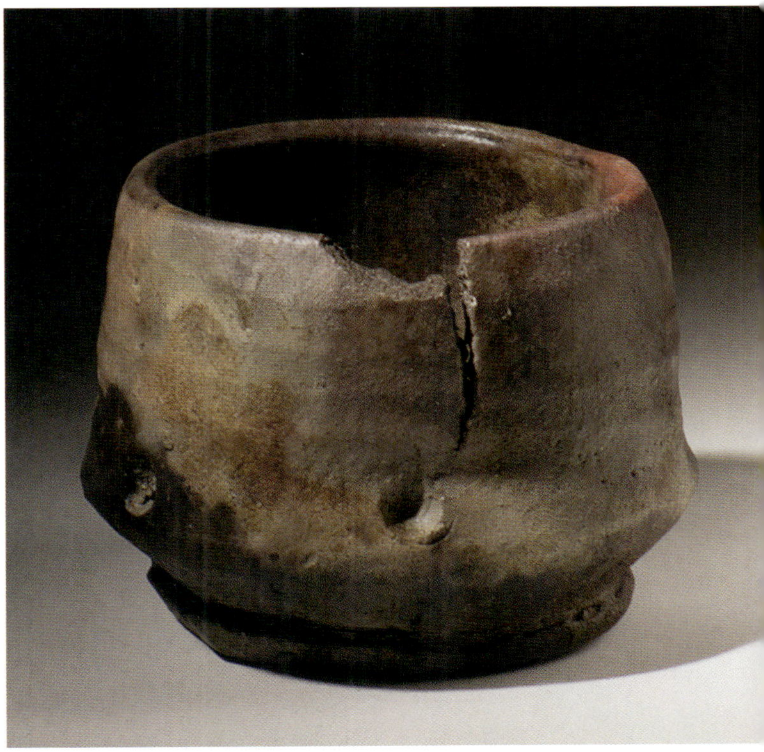

Peter Voulkos. *Untitled Tea Bowl*, 1998, stoneware, pit fired, 5" x 6" x 6 ¼". *Private collection. Photo credit: © schopplein.com. Courtesy of the Voulkos & Co. Catalogue Project.*

PIT FIRING IN AUSTRALIA

Unlike America, where, as we have seen, artists were able to draw from an indigenous pottery history, Australia had no pre-existing pottery tradition to explore. The Aboriginal people of Australia were largely nomadic, and did not develop agricultural practices, and therefore had no need to build strong, permanent structures. Technologies and skills such as brick-making and tiling, therefore, did not develop. Being nomadic, carrying heavy ceramic implements was burdensome; hence, permanent, fired domestic wares were neither needed nor developed. Further, the cultural conservatism of the Aboriginal people (Flood 2006, 8), due to the continent being so large and geographically isolated, meant that cultural cross pollination did not occur; therefore, ceramic knowledge was not introduced to the continent. Much of Australia's early ceramic history therefore was imported, mainly as an English influence, during colonial white settlement.

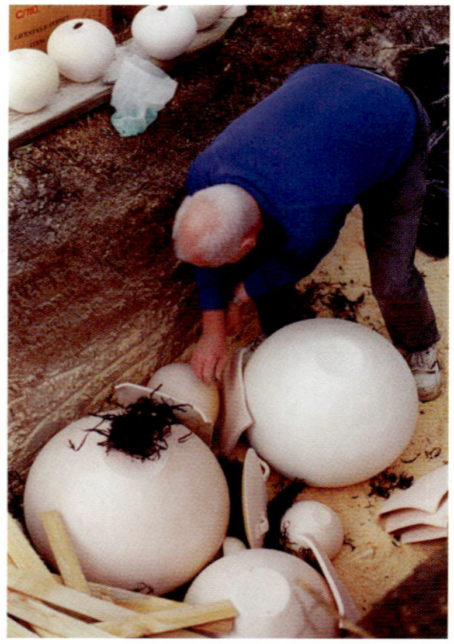

Ray Rogers loading his pit: a layer of sawdust is spread, into which the pots are nestled. Salt, copper carbonate, iron, and seaweed are sprinkled around pots, and the pots are covered with a layer of wood, then the pit is set alight. The pit is covered with corrugated iron. The firing takes five to six hours and twenty-four to cool. *Photo: Ray Rogers*

As in America, the post-WWII years in Australia ushered in a period of prosperity and increased leisure time, and the trend of experienced potters taking up teaching positions in art courses followed the American model. Whereas Australia had traditionally looked to England for social and political guidance, America stepping in to "win" WWII had a snowball effect on their political and social influence throughout the world, so that, aided by improvements in mass communication, the radical changes in the American art world began to filter through to Australia. As in America, artists began to use clay as a sculptural medium to communicate social and political statements, and the long reigning Leach/Homada tradition, which had dominated Australian ceramics slowly began to peter out as new attitudes toward function and clay as a fine art material arose.

Pit firing was introduced to Australia by New Zealander Ray Rogers in 1983, after he moved to Australia to undertake a nine-month residency with the Potters Society of Australia (Lockwood 1984, 29). Rogers discovered the technique in California in 1980, when he joined a group of students and potters conducting a low-firing workshop on a beach in San Jose. The event took place over four days, and included participants making work and kilns for a range of low fired techniques, including wood, raku, and pit-firing (Lockwood 1984, 30). Rogers enjoyed the spontaneity of the fire and the resultant intergalactic patterns that swirled across his pots. Upon returning to New Zealand in 1981, Rogers was given a grant by Queen Elizabeth II (QEII), New Zealand, to develop pit firing (Cooke, cited May 14, 2010). Prior to this, Rogers had been influenced by the Leach/Hamada philosophy, as were many potters in Australia and New Zealand at the time, and was mainly making high fire stoneware functional work. A highlight in Rogers career occurred in 1997 when he demonstrated his pit firing technique at the International Potters Festival at Aberystwyth in Wales. His forms are mostly simple, spherical forms aimed at heightening the subtle, yet dramatic, effect of the open flame on the ceramic surface, though he has also designed and fired tiles for murals. Some of these wall panels have been purchased for corporate buildings in New Zealand, including the Sheraton Hotel in Auckland (Rogers, 1983, 23).

Since Rogers introduction of this technique to Australia, many ceramicists have incorporated this method in their practice, one of the earliest being Barry Hayes, who recognizes "a strong tradition in pottery, an unbroken chain that links us to the earliest potters" (Hoare 1987, 30). As a young boy, Hayes lived near a brickworks that used a wood kiln to fire the bricks: this kindled his interest in

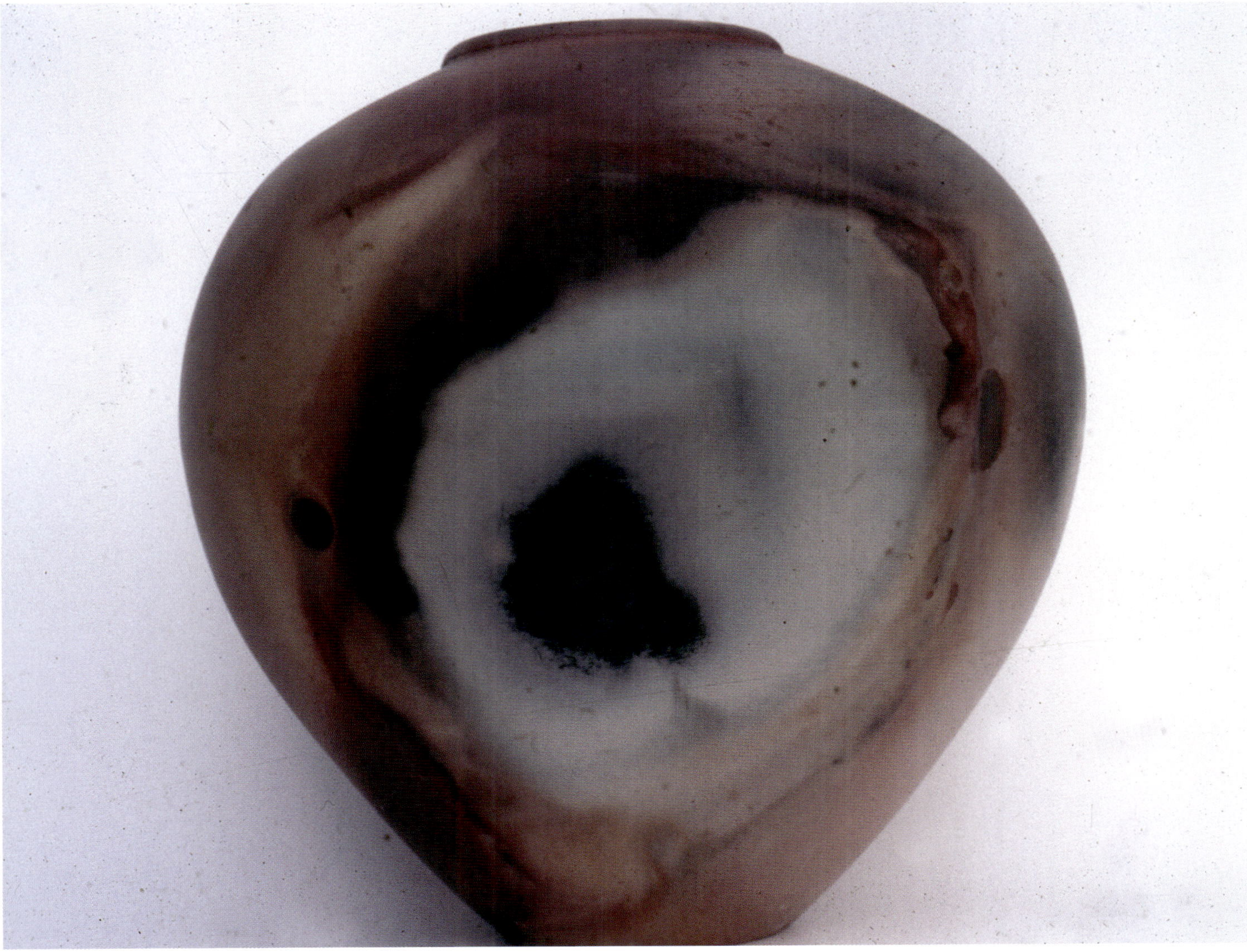

Barry Hayes. *Vase form*, 10" x 8". *Photo: Barry Hayes*

fire. At nineteen years of age, Hayes went to Papua New Guinea (PNG) to work and discovered, and fell in love with, indigenous traditional pots and the methods of making them. Upon returning to Australia, Hayes spent six years in two production potteries, where he faithfully and repetitively reproduced shapes, glazes, and decoration. The monotony of this experience reignited the excitement he had felt in PNG and fuelled a desire to rebel against the tedium of production pottery. After leaving the potteries, Hayes began pit firing, and continued to do so for the following twenty-five years. It is easy to see why pit firing harmonizes with Hayes's philosophy and art practice.

CHAPTER THREE
HOW-TO

[M]inerals leave me cold. For that reason I never make glazes. Glass cuts, it is rigid. It lacks life. I never manage to be moved by it despite its beauty (Bayle in Perryman 1995, 6).

COLOR PALETTE CHART

Organic and non-organic materials used in pit firings.

Organic Combustibles, Chemucals and wires, etc	Possible Color Effects (depending on pit temperature/atmosphere, etc)
Hardwood sawdust	Black, dark gray
Hardwood coals	Black, smoky gray, blue-grays
Cow pats	Black, grey, golden yellows
Seaweeds	Browns, rust, honey, orange, peach
Salt soaked straw	Warm orange, peach, muted yellow blushes
Salt	Warm orange, peach, muted yellow
Copper carbonate	Green, black, maroon, reds
Cobalt carbonate	Blues
Red iron oxide	Warm orange
Ferric chloride	Iron reds, yellows, oranges
Copper wire	Green, black, maroon or red lines
Salt soaked string	Warm orange, peach, muted yellow blushes lines
Copper scourers	Green, black or grey fumed patterns
Chicken wire	Reief fumed patterns

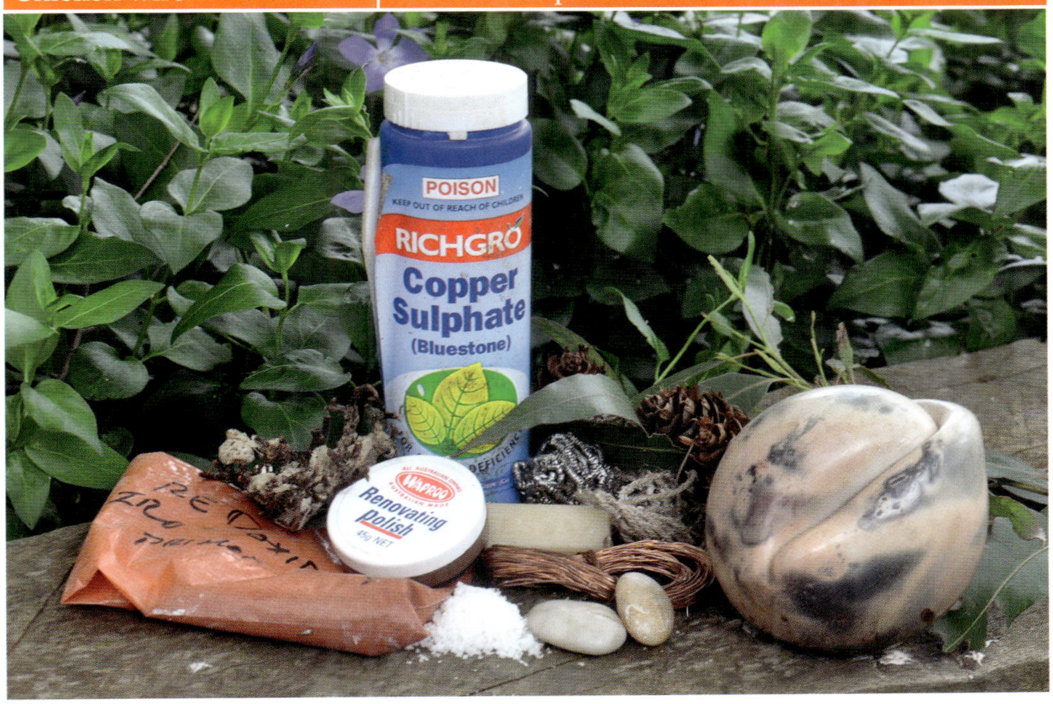

Contemporary methods employed for pit firing can fall into two categories: those used by the traditional potters of India, Pakistan, South America, and North America, who continue to use this technique to fire their unglazed functional wares, and those used by ceramicists, who use the technique for the artistic outcomes they are able to achieve on their artworks. We have seen in the previous chapter the precise and calculated methods used by traditional potters to achieve the surface finish desired for their functional and decorative wares: burnishing is employed, bark, wood, and/or dung is used for fuel and reduction or oxidation atmospheres are regulated in the pit depending on the desired color outcomes. Most ceramicists who use this method to further the artistic expression of their works, however, introduce many more materials in their pit firing regime, both organic and non-organic: seaweeds, various plants, dung, copper wire, steel mesh, metallic oxides: the list is considerable.

Despite these differences, a technique in common that many ancient, traditional, and contemporary potters use in preparing their pots is burnishing, because not only does the subtle sheen achieved highlight the organic tonal hues attained by pit firing, but many pit firers say they enjoy the meditative state induced through burnishing.

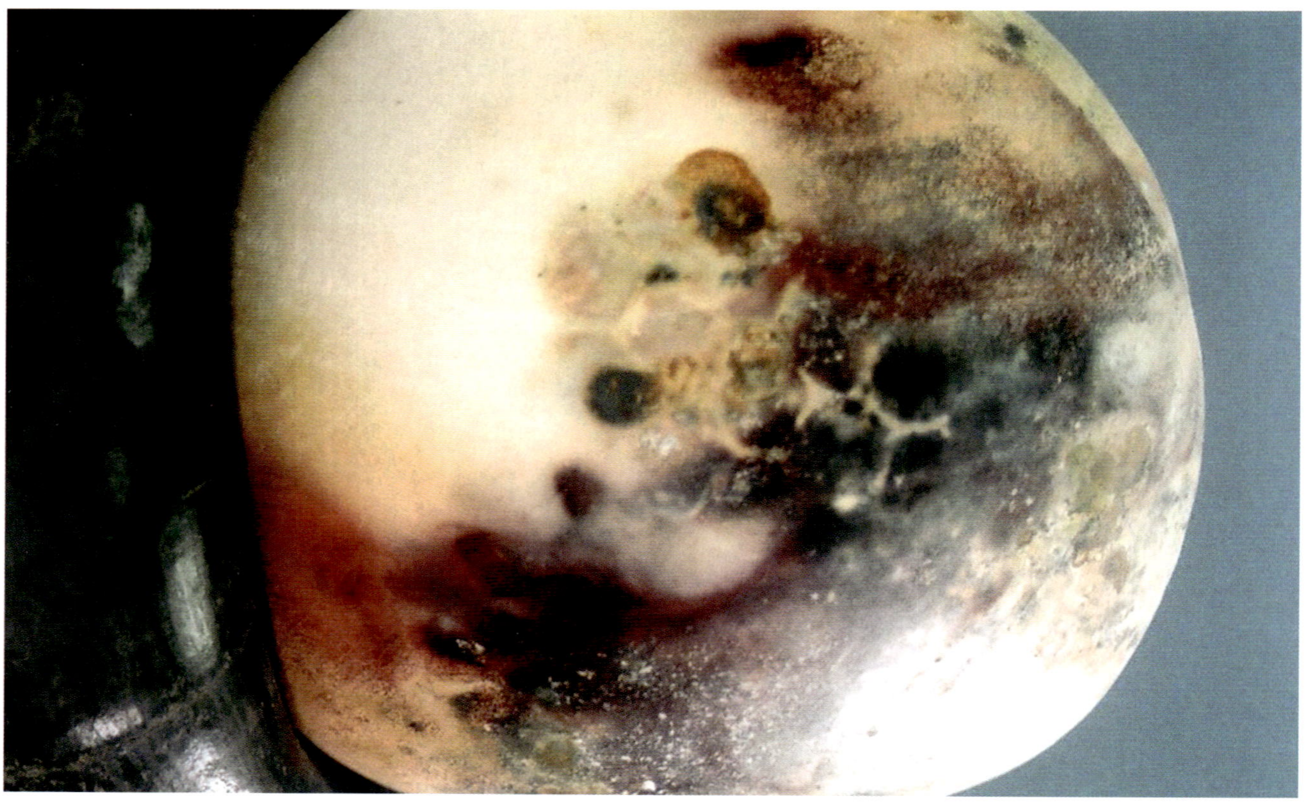

Dawn Whitehand. *Nestled* showing fumed chicken wire—chicken wire is wrapped around the pot as tightly as possible, so that wire remains in contact with the pot during the firing. *Photo: Author*

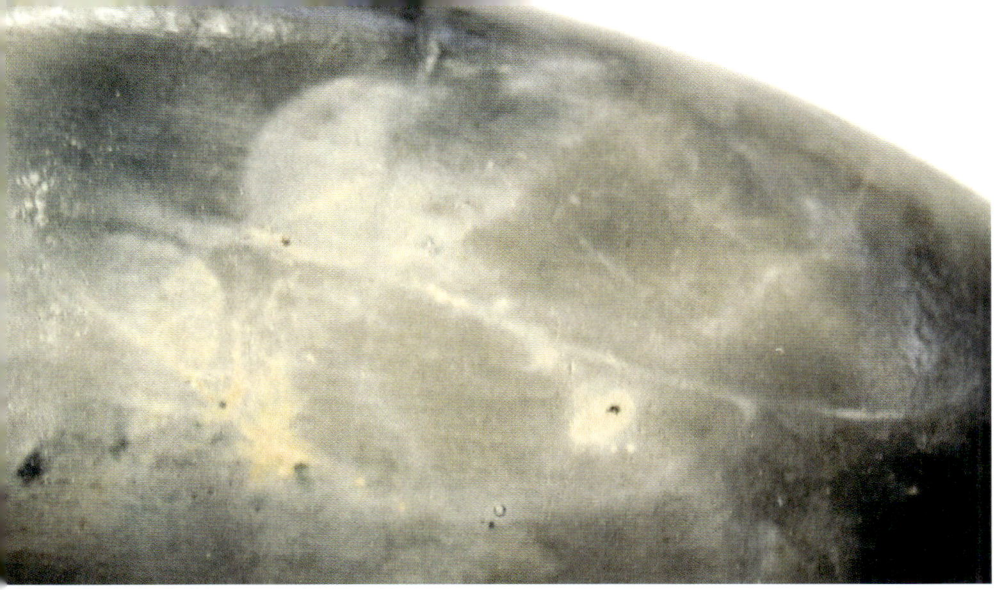

Dawn Whitehand. *Starry Night.* Shows fumed gum leaves—gum leaves are placed on sawdust and a pot is placed on top. *Photo: Author*

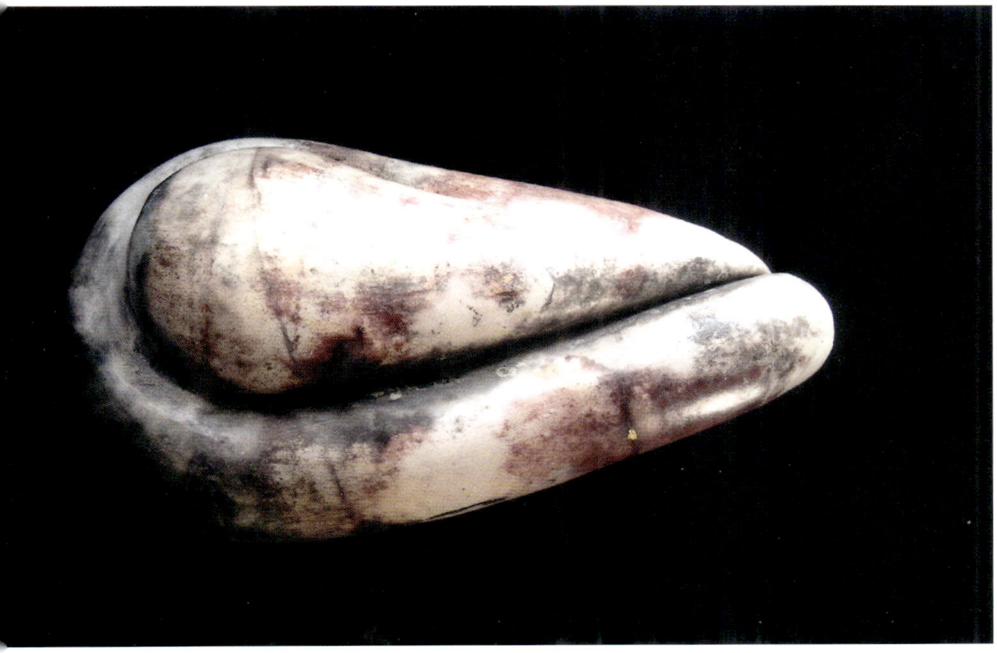

Dawn Whitehand. *Voluptuous Earth* showing black lines left by copper wire wound around the clay body. *Photo: Author*

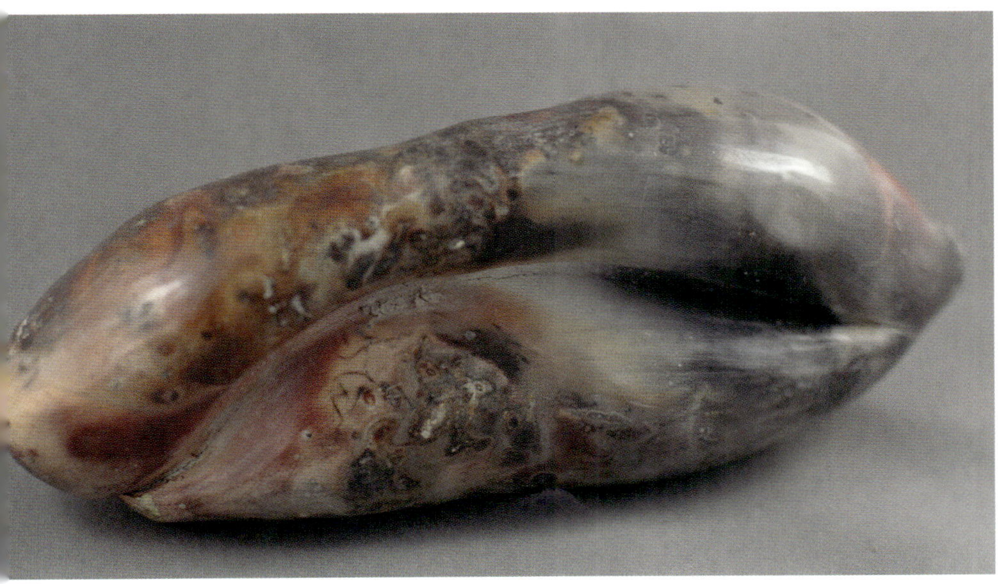

Dawn Whitehand. *Landscape.* Shows blackened surface when pot is buried in sawdust. The colored area was wrapped with fine seaweed and the patterns of the seaweed strands can be seen as wavy lines.

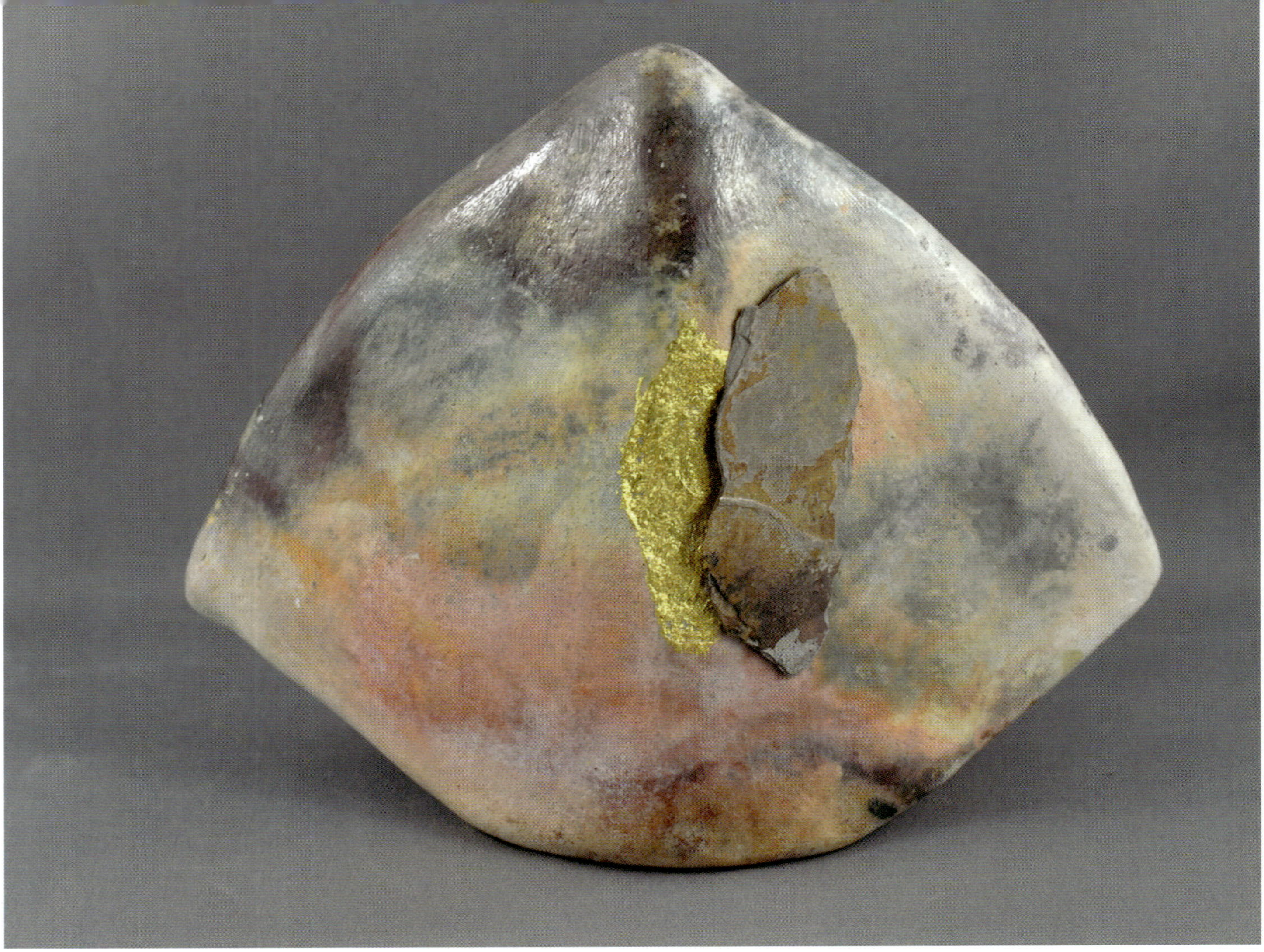

Dawn Whitehand. *Volcano.* Pit-fired wrapped in salt-soaked straw leaving a soft peach hue. *Photo: Author*

Dawn Whitehand. Burnished greenware showing crevices after manipulation. *Photo: Author*

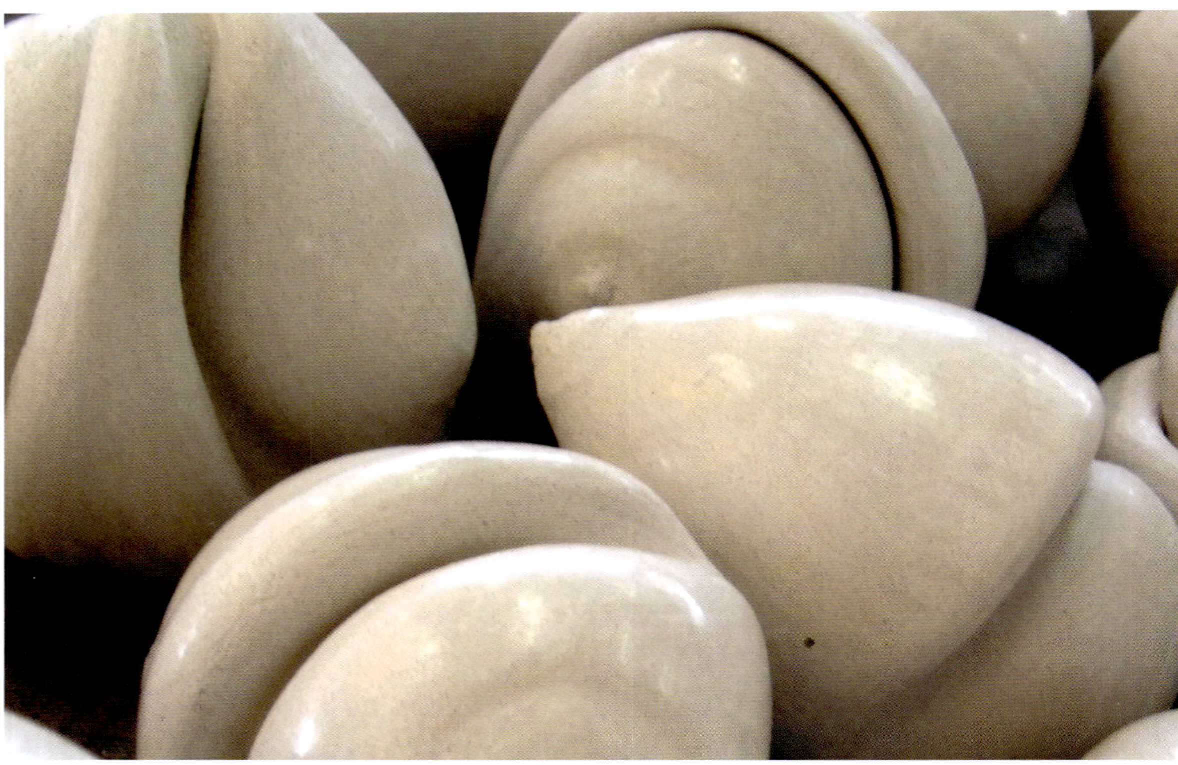

BURNISHING

The Potter's Book, mentioned earlier, contains an Islamic prayer for the "smoothing of pottery with a stone," (Rye 1976, 192) or what we refer to as burnishing, indicating that early Pakistani pottery was also burnished before being placed in the pit.

Many early forms exhibit the use of burnishing, and may have developed from the use of the technique on the floors and walls of residences, probably to strengthen and waterproof them, in places such as Neolithic Catal Hüyük (Mellaart 1965, 220). Again, this practice seems to have developed independently around the world, and was probably applied to pots for both aesthetic purposes and the need to have a smoother, more water resistant cooking surface. The excavation of the Cupisnique kilns (ca. 1000–700 BC) on the north coast of Peru revealed the presence of worn hematite, suggesting they were used as polishing stones (Shimada 1997, 110). Similarly, amongst the artifacts found at the Provincial Chimu kilns (ca. AD 1375–1470) in Pampa de los Burros, were stone burnishers (Shimada 1997, 121).

Today we understand that burnishing compresses and realigns the clay particles, enabling the strengthening and waterproofing qualities desired by early civilizations. Many contemporary artists exploit this technique, burnishing their greenware up to three to four times in order to achieve a highly marble-like, reflective surface. The resultant sheen emphasizes the subtle tonal variations achieved by the modern pit firing method.

When burnishing my works, I use a smooth river stone that fits my hand perfectly. Burnishing is time consuming and repetitive, but using a stone that fits into my palm means I am less likely to get cramps in my hands and fingers. Depending on the result of the first burnish, I may do a second, and then allow the piece to dry. I then rub baby oil into the surface of the piece and reburnish to a high sheen. Works are then low bisque-fired to 1500°F. Thus, the burnished sheen remains on the pot, renders the pot strong enough to handle during preparation and loading, and provides enough strength against thermal shock in the pit.

There are as many ways to make, prepare, and fire ceramic works as there are ceramicists, and some of these methods are featured on the following pages.

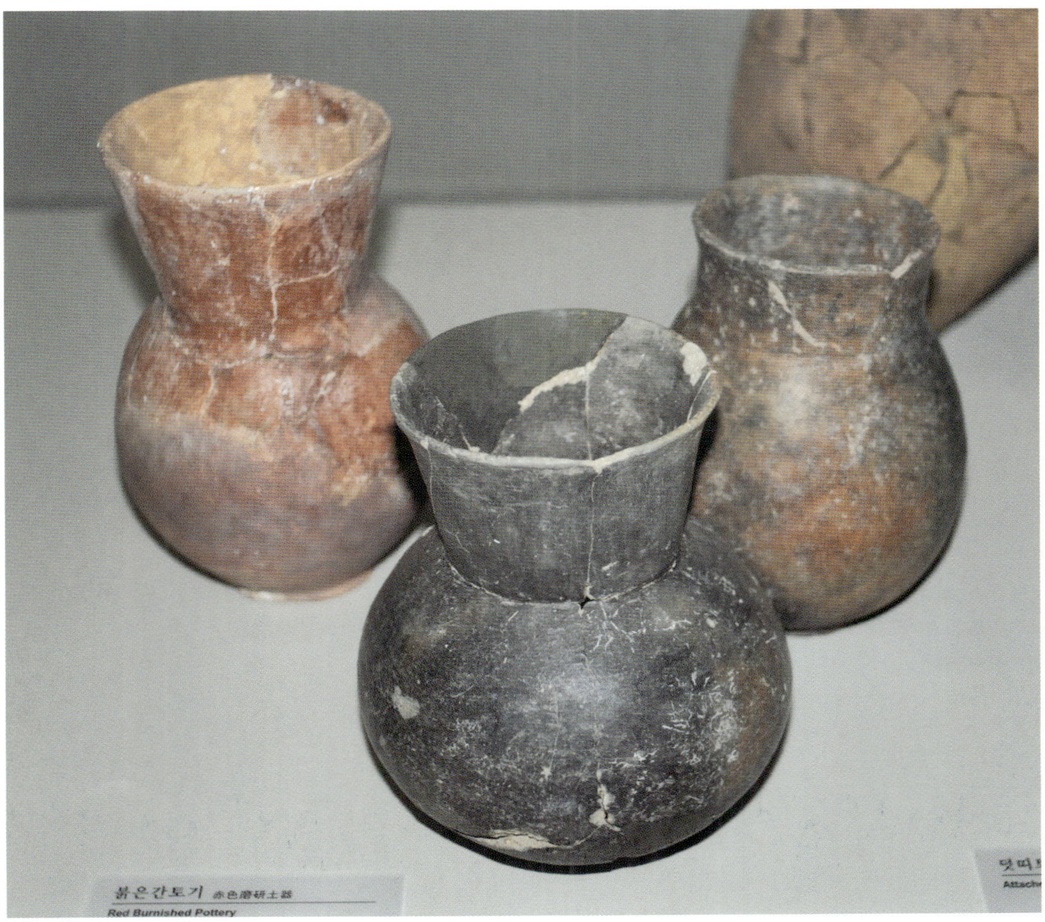

Burnished red pottery, Bronze Age and black burnished pottery, Early Iron Age. *Collection: Gimhae National Museum, South Korea. Photo: Author*

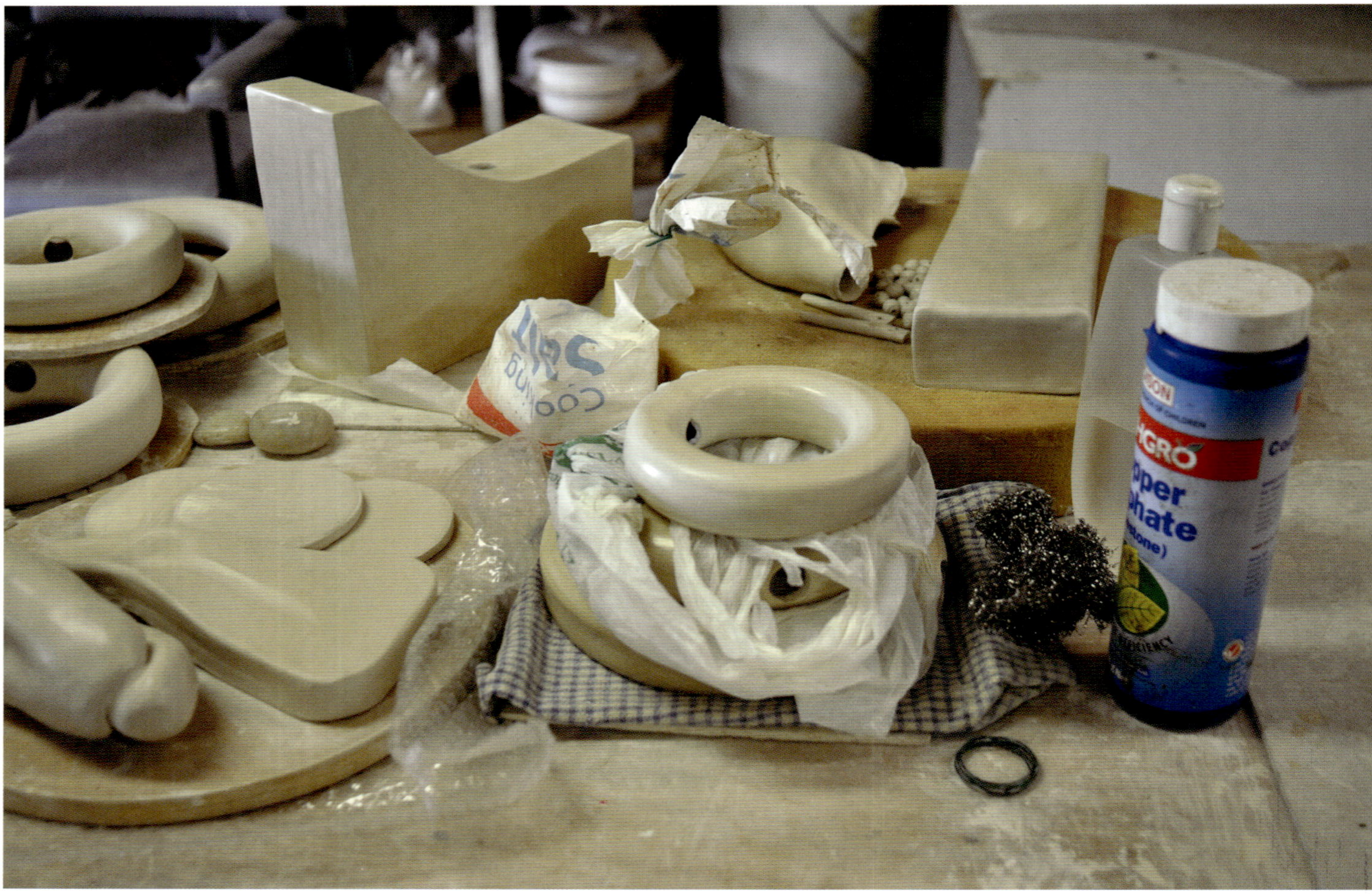

In the studio: preparing for a pit firing. *Photo: Author*

Dawn Whitehand

I throw and manipulate my ceramic sculptures using a grogged stoneware, preferably a clay with smaller, rounded particles that is less harsh on skin during the throwing process and burnishes without scratching. I use a method I call double ended throwing: after centering, the clay is completely opened through to the wheelhead; the vessel walls are raised, leaving the lower walls thick, and the top is completely enclosed so that there is no opening. After removing the form from the wheel, the lower half is wrapped in plastic, so that it remains soft, while the top half is allowed to dry to medium leather-hard. Timing here is crucial. If this top section is too soft, it will collapse over the sides of the chuck; if too hard, it would not be possible to manipulate the finished form. When ready, the form is inverted into a chuck on the wheelhead, so that the lower half is now the top half, and it is still soft enough for throwing. The leather-hard state of what is now the lower half provides enough strength to throw the soft end of the form. This end is also thrown to be fully enclosed. The end result is an elongated egg-like shape. The end in the chuck is now sprayed lightly with water and covered lightly with plastic so that the entire piece can attain an even, soft leather-hard state in preparation for manipulation. Folds and creases are created in the pieces with my fingers, an intuitive and tactile process.

When the pieces are leather-hard, I begin the burnishing process, which has been described. The sculptures are then low bisque fired to 1500°F to retain the burnished surface and to allow the clay body to remain porous enough to absorb the fuming within the pit. After bisque firing, the pieces are prepared for the pit. Individual works are wrapped in combinations of seaweed, gumleaves, copper wire, chicken wire, and salted string.

Next, the pit is prepared. Most of the ingredients used in the pit are readily available on my property, making this aspect of my art practice more locally sustainable. The pit is prepared by layering wood on the bottom and the sides, eliminating as many cold spots as possible. Layers of oyster shells, used for their mineral and salt content, together with pine cones and cow dung are added. These items generate a solid, lasting heat, especially the cow dung, whose thick ash can also provide areas of reduction during the firing. The forms are then placed on the bed of organic material in the pit, and any spaces between the works are filled with more oyster shells, pine cones, and sawdust, allowing for even combustion. Copper sulphate, salt, iron oxide, and epsom salts are sprinkled liberally around the forms. More dung and pine cones are then layered over the work, and other combustible material, such as dry leaves and wood is loaded on top.

The pit is then set alight. It is periodically stoked for two to three hours to build up the embers. Once an even glow is achieved, a final layer of wood is added and the pit is covered with corrugated iron, restricting oxygen and allowing for a slow combustion firing. The pit is usually cool enough to open and unload 24–36 hours later. While I may be able to reasonably predict the outcomes of the coloration on the forms, there is always an element of randomness to the pit firing process, depending, for example, on the quality of the seaweed, the type of wood used in the firing, how hot the pit becomes, and how cold the ground remains.

When the work is removed from the pit, it is covered with a fine layer of ash, which is removed with a damp cloth or soft brush. Works are not washed under running water, as the surface is still porous and any surface salts still on the form may become unstable, triggering a gradual corroding process. There are many methods that can be utilized to retain the integrity of the delicate low-fired surface. Beeswax and neutral shoe polish both achieve a subtle sheen, gently highlighting the soft coloration in keeping with the natural forms and soft appearance of the forms.

Rob Drexel

Rob Drexel's pots are wheel-thrown with hand-built additions. He paints the surface of his green vessels with terra sigillata that is specially prepared from high iron content clay and a technique that he has developed over the years to be particularly reactive in the variable oxidation–reduction environment he creates in his firings. Drexel describes his process: the bisque is done to cone 012—about 1600°F. This leaves the vessel porous and able to soak up the vaporized salt and copper carbonate in the pit. It is extremely important to use fine salt and get the fire hot enough to vaporize that salt, breaking the chemical bonds of the NaCl and liberating the sodium to bond to the pottery surface as that wonderful rich yellow color. That requires 1500°F or more—a good hot bonfire! This liberates the yellows from the sodium ions and the copper ions from the copper carbonate, enabling them to blend in the pit to create a wide and beautiful range of bright yellow to orange, to maroon, and finally a deep red. The blacks come from the sawdust lining and smoldering against the side of the pottery piece at the bottom of the pit. The pit should be held in slight reduction as the fire passes its bonfire-like peak by covering it just enough, but not too much as to smother the fire. This keeps the sodium and copper from re-oxidizing and lets them accumulate on the pottery surfaces with all the variations that the irregular burning in the pit provides. After removal from the pit, Drexel cleans his works using a brush, and varnishes the surface as it dramatically heightens the colors and creates a washable surface for people to keep the piece looking its best—after all, according to Drexel, it is decorative and not functional.

Suzanne Schmaltz

Suzanne Schmaltz prefers hand building over other forming techniques. The methods that she uses in the formation of her ceramic bases lend themselves to a final finish that has a leather-like quality. Schmaltz enjoys earthy, roughly finished ceramics. Her primary tool is a slab roller and she uses a slump method for her bases, draping clay sheets over a variety of shapes and working the vessel upside down to achieve the desired form. She leaves the edges rough and unfinished, often impressing leaves or other textures into the clay surface. At the near leather-hard stage, holes are punched around the upper edge for the future incorporation of pine needle basketry.

Schmaltz bisque fires her ceramics in an electric kiln before the pit firing. Although she experiments with various chemical combinations and firing materials, Schmaltz prefers red and yellow oxides as well as copper carbonate, Miracle Grow fertilizer, and coarse sea salt; she favors the markings from the salt. The pit is prepared using a layering process of equal amounts of wood shavings and hay, onto which generous amounts of chemicals and salt are sprinkled. Pots are nestled

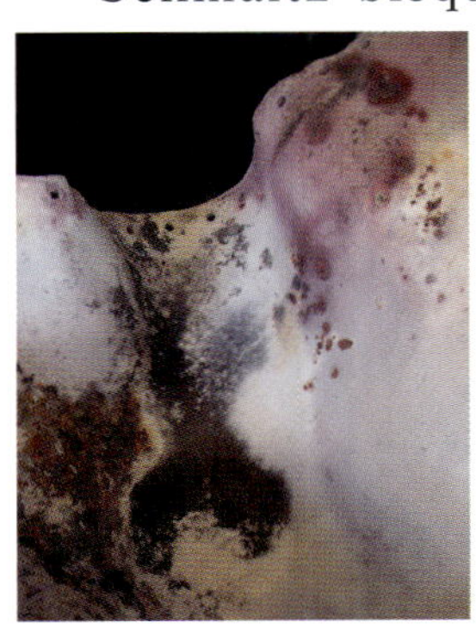

Suzanne Schmaltz. Work in progress showing holes for later incorporation of pine needle basketry. *Photo: Suzanne Schmaltz*

into this mixture and then topped with another layer of sawdust and hay and additional chemicals. A second layer of ceramics with shavings/hay/chemicals is sometimes added, but never more than two ceramic layers. Over this wood of various sizes is laid for maximum heat generation. An initial hot fire begins, adding more wood as necessary, and is allowed to burn for about a half an hour, after which a heavy, vented steel cover is placed over the pit to let the fire burn and smolder overnight. Firing time is usually four to eight hours.

According to Schmaltz, the best part of the entire process is uncovering the pit and discovering the uniqueness of coloring and markings on each piece. "It always feels like Christmas morning and discovering what Santa has left under the tree."

Michael Wein

Michael Wein's work is wheel-thrown. Pointed pots, such as *Untitled 2* (image in Artist Profile section), are thrown upside down. At the leather-hard stage, Wein burnishes the pots with a glass candle light bulb, which only lasts about five to six pot burnishes before wearing out. All pots are burnished three times: one—left to right, two—right to left, and three—up and down, with stokes only going in one direction, followed by a final burnish with his finger so no that ridging is visible on the burnished pot.

When dry, Wein's pots look like polished marble. It can take Wein almost a day to finish one large pot. The coral or textured pots are carefully carved at the leather-hard stage. Wein lives on the beach at Nelson Bay in New South Wales, Australia, and is inspired by the corals and seaweeds he sees along the coastline. When he pit fires seaweed pots, he partially wraps them with thin Kaowool insulation to keep the seaweed in contact with the pot so he can achieve specific markings. Sometimes he soaks the seaweed in a brine of copper carbonate and salt. Packing a pit is most important as the parts of pots in sawdust will be black. Wein places salt and oxides as close to the pots as possible, but not touching the pots, and hopes for a good firing. But, most of all, says Wein, he has a good time experimenting.

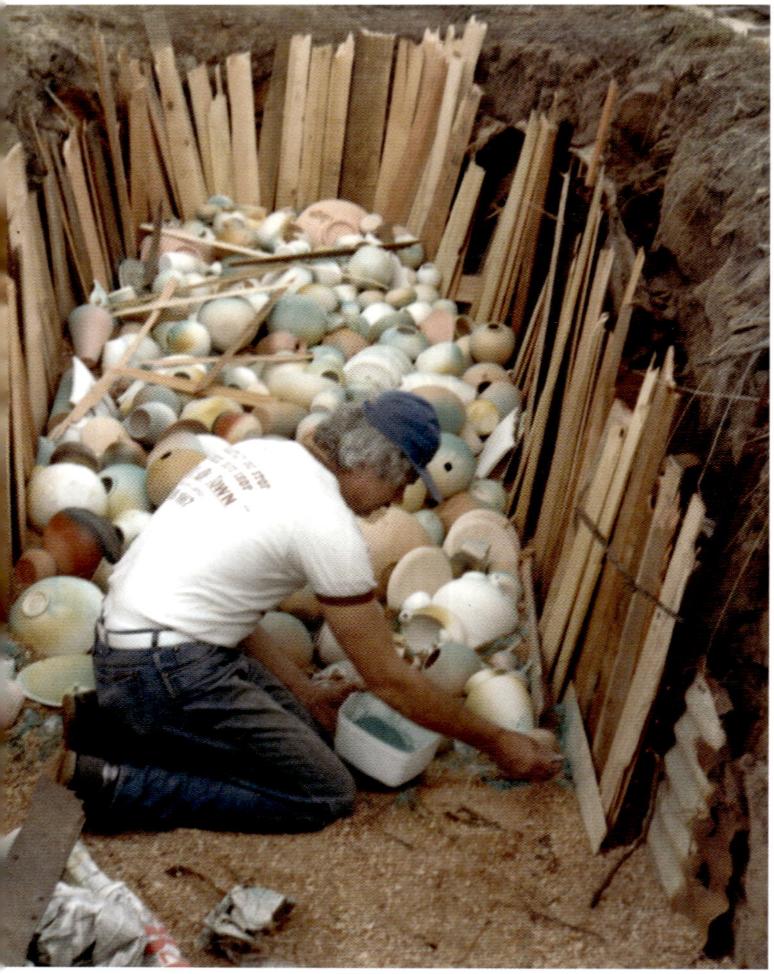

Michael Wein loading the pit and sprinkling copper carbonate. *Photo: Robyn Wein*

Keven Francis

Keven Francis has experimented for many years with pit firing, and recently developed a method he calls Dynamic Pit Firing, so called because he is continuously and directly engaged in the firing process. Francis begins with palm-sized pinch pots,

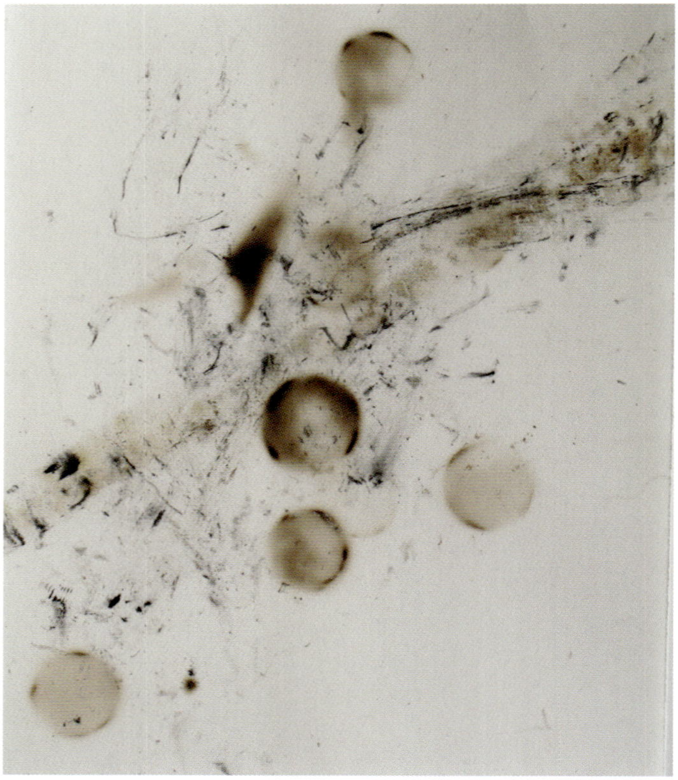

Keven Francis. *Impressions Three.* Scorch drawings. *Photo: Keven Francis*

which he likes for their intimate yet robust quality, burnishes them so as to leave faint ripple lines around them accentuate their tactile properties, and bisque fires them to 1290°F. The bisque pots are then placed on sawdust in the pit and covered with layers of soft and hardwood. Once the sawdust is lit, Francis controls the fire by lifting and changing the pit cover, accelerating or slowing the fire, and reaching into the fire pit to move burning timber and heated pots. The firings are fast and intense, with the ceramic pieces and burning timber manipulated to infuse and cut away lines of carbon. Within forty-five minutes, the pit is set, fired, and the pots removed. Looking for a way to extend this direct interaction with the fire, Francis begins to remove the hot pots, placing them on paper together with embers and charcoal. The resultant scorch marks comprise an indexical drawing that echoes the firing event, in images from pots and charcoal (Francis 2010, 73–4). In essence, a record of the dynamic, yet ephemeral firing process.

Bennett Bean

American Bennett Bean's process is different, not so much for his pit firing method, but because of all the other elements he incorporates into his pit fired artworks. After throwing and burnishing a white earthenware body, Bennett alters and adds slab elements to the form. The piece is then painted with terra sigillata and polished. After bisque firing stencils and graphic tape are applied to the piece as a relief and glazed, after which pieces are pit fired. After pit firing, the insides of the pieces are prepared for gilding with gold leaf. Finally, the exterior is painted and the completed ceramic sculpture is sealed with urethane gel.

Cathy Jo Buhlert

I was aware for many years that the pit firing ceremony would be the most appropriate firing method for my work, but it was only when I moved out into the desert of southern Arizona that everything fell into place to begin the grand experiment. I was living in a small travel trailer on a ranch belonging to a friend at the time. I had been making pots for weeks and there was no kiln in sight. I had never pit fired nor seen anyone else pit fire, so I was apprehensive to say the least. I went out to walk and pray in the desert and visit an old grandfather saguaro … on the way back I heard a Spirit say, "All you lack is experience." I knew it was the answer to my prayers. There was only one way out. I would be required to move forward on faith. My first pit firing was a very surprising success! My second pit firing showed me in no uncertain terms the potential this low, simple technology held.

Cathy Jo Buhlert. View of the freshly fired pit.
Photo: Kathy Guellich

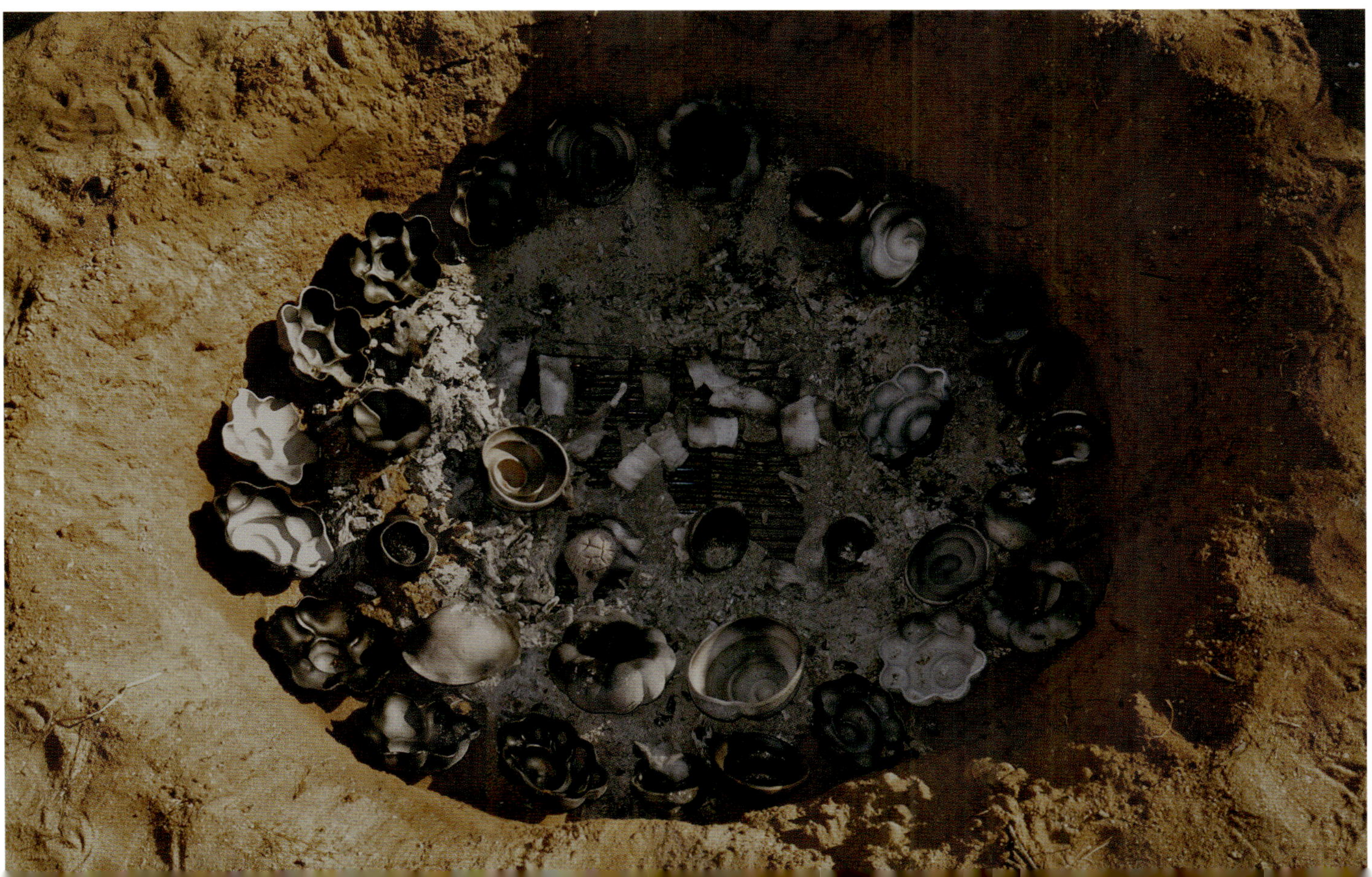

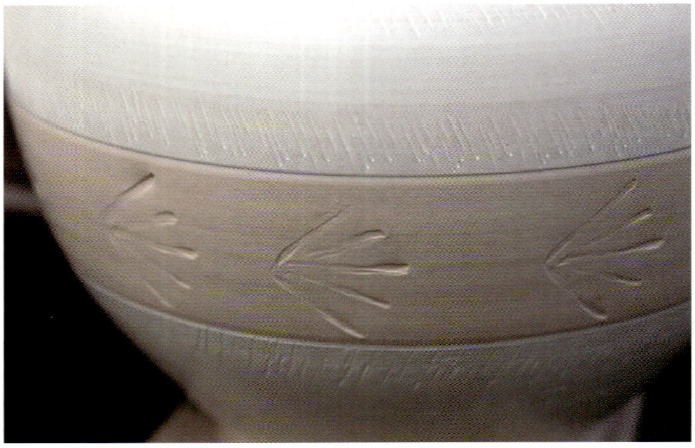

Genez Malebranche. Close up of Terra Sigillata and bare clay body before firing. *Photo: Genez Malebranche*

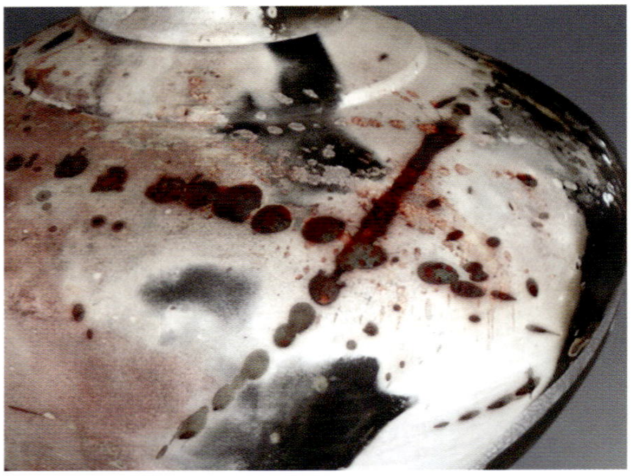

Genez Malebranche. Detail of raku splash on *X Marks the Spot*. *Photo: Genez Malebranche*

I fire my pots in a very peaceful and beautiful location. The land is populated with older mesquite and provides an excellent supply of fuel. I use small sticks to feed the hungry fire, slowly, carefully heating the raw pots. Necessity truly is the mother of invention. I use the sand, dug out to create the firing pit, to bury the pots for reduction. Like a mother turtle burying her eggs, I cover the pots with this sand, creating a safe nest as they "incubate" in the hot coals. I fire my pottery entirely by intuition; the "experience" I need will come with the years. The ceremonial nature of my firing process is essential to my work.

Genez Malebranche

In an attempt to add a twist to her artwork, Genez Malebranche has developed a unique approach to pit firing. While Malebranche employs many traditional pit firing methods, when it comes to the actual firing of the works, her pre-pot preparation is sometimes a little different. Many ceramicists who paint terra sigillata on their works cover the entire form. Malebranche, however, only partially covers her leather-hard pieces, leaving some areas naked, because the "terra sigillata absorbs the colors of the pit differently than the bare clay, therefore creating drama from the different effects on the same pot."

The other technique Malebranche uses to introduce "drama" to her artwork is a technique she calls Pit Fired Raku. Malebranche brushes or splashes raku glaze on dry greenware and then bisque fires the works to cone 06. They emerge from the electric kiln with none of the expected reduction that comes with Raku methods, but that will develop in the pit. The Raku is applied sparingly through fast, spontaneous movements to obtain variation of surface during firing. Glazing the entire form would render the process pointless—the work may as well be Raku fired!

When the finished artworks have emerged triumphantly from the pit, Malebranche gently cleans them so as not to scratch the delicate surface. The sealant she uses to protect the subtly colored pieces depends on whether they have been burnished or not. Bowling alley wax is applied to burnished works because it polishes to a lustrous satin shine that gives a spectacular three dimensional quality to the surface of the artwork. This wax should not be applied to unburnished or heavily textured pieces, as the wax will build up in the texture and be very difficult to remove. Instead, Malebranche uses tile and grout sealer for non-burnished works because it takes on the texture and finish of the surface being sealed. So, matte areas still look matte and shiny areas still look shiny, while also protecting the work from grease and possible staining.

Alex Mandli

Each piece is thrown on the potter's wheel from white earthenware clay that matures at (cone 04) 2,008 degrees Fahrenheit. Once the piece is leather-hard, a polished agate stone is used to "rough burnish" the surface of the piece. This mashes down the larger particles of clay, smooths the surface, and brings up the finer clay particles.

The piece is then dried for a few hours, and then the final burnishing is started. This final burnishing makes the surface extremely smooth and shiny. There is no glaze applied to these pots.

When the piece is bone dry, three to five coats of terra sigillata (very fine particles of clay) are sprayed onto the pots and polished with the palm of the hand and a chamois (soft leather). The work is then bisque fired at a relatively low temperature (cone 016) 1,517 degrees Fahrenheit. Bisquing to this temperature gives the work a fair amount of durability. At this point, the pieces are bright white.

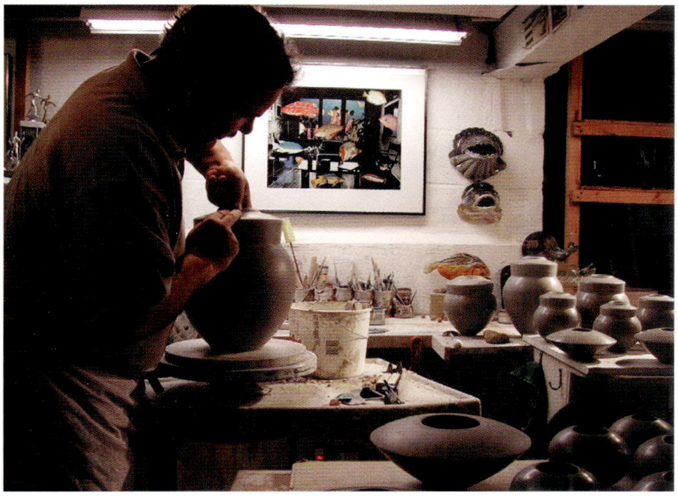

Alex Mandli. Throwing the pots. *Photo: Jean Mandli*

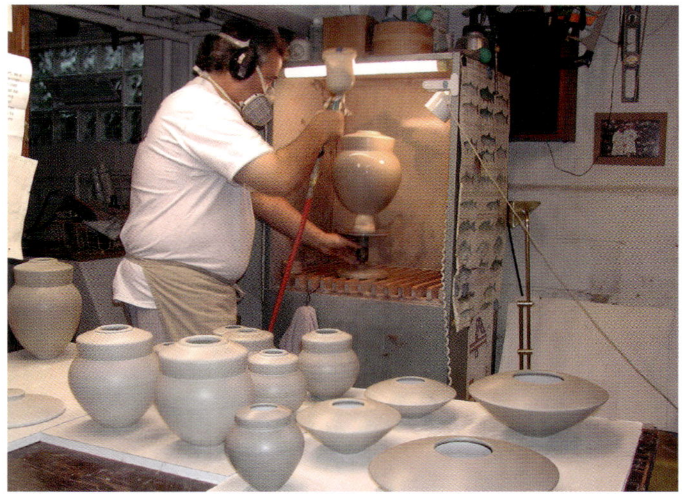

Alex Mandli. Applying Terra Sigillata. *Photo: Jean Mandli*

The final step is a firing technique that many ancient cultures once used. The work is loaded into a pit (a 3' x 4' hole in the ground), and fired in combustible materials (fuel). These materials may include burled wood shavings, a variety of sawdust, newspaper, straw, metal shavings, ceramic frits, used steel wool and sandpaper, and manure. The fuel is essentially someone else's garbage.

The burning of these materials traps carbon on the surface of the burnished forms. Where the fire burns hot, the surface of the form will be gray to white. Where the fire burns slowly, it will be black. The firing time is between eighteen to twenty-four hours. Once cooled, the finished pieces are cleaned and polished three more times.

Tania Rustage

Tania Rustage throws her pots on the wheel, burnishes them, and applies terra sigilatta while rotating the pots on the wheel. They are then buffed to a high shine before being bisqued at cone 010. In addition to terra sigillata, Rustage has also experimented with a coating of 50/50 copper carbonate and Miracle Gro, but has yet to see any significant results. She uses a variety of organic, combustible material in her pit, including wood shavings, horse manure, coffee, salt, seaweed, corn husks (soaked in salt), and pine needles. The pots are nestled amongst and beneath these materials and then layers of wood and logs—which are sourced while Rustage is on fishing trips with her husband—are placed on top. Once the pit is alight, Rustage throws a mix of wood shavings and copper into the fire so that the copper does not actually touch the surface of the pots, but fumes it, so that a crust will not form on the surface.

The following day the pit is cool enough to reveal the treasures.

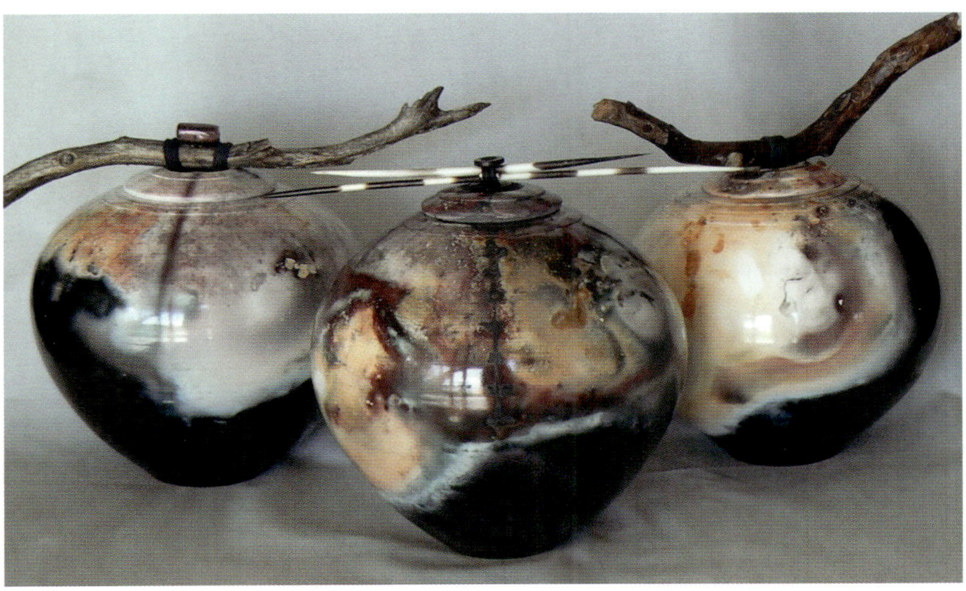

Tania Rustage. A grouping of cleaned, waxed, finished pots with wood additions. *Photo: Tania Rustage*

PIT FIRING
PHOTO ESSAYS

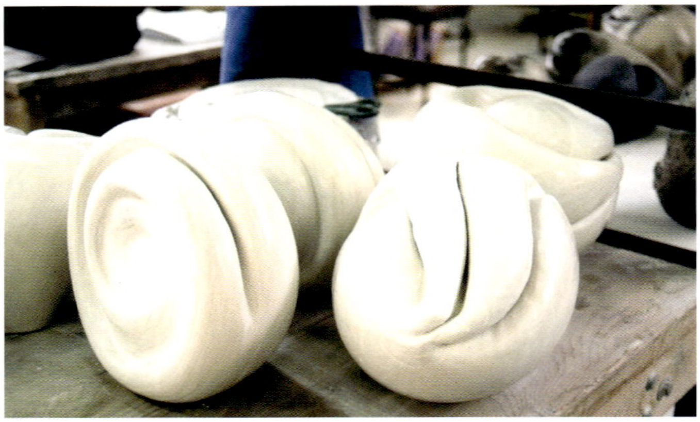

Burnished and low bisque fired ceramics ready to be prepared for the pit. *Photo: Author*

Dawn Whitehand

I fire in a pit that I adjust in size with a wall of bricks depending on the amount of work to be fired. The pit measures 51" w x 83" l x 36" d. My firing process is described in the previous section.

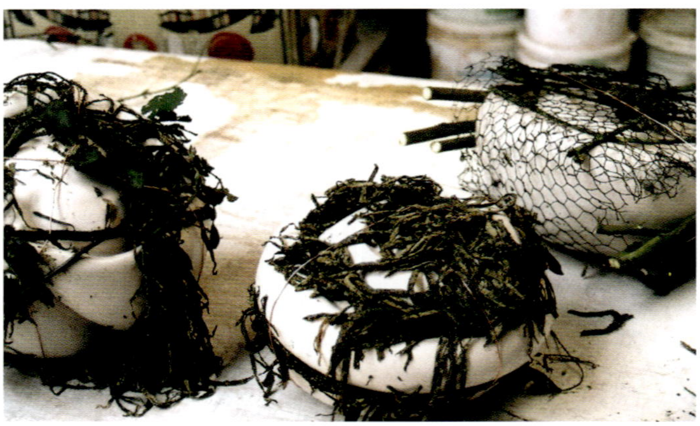

Ceramics wrapped in seaweed, leaves, copper wire, and salted string. *Photo: Author*

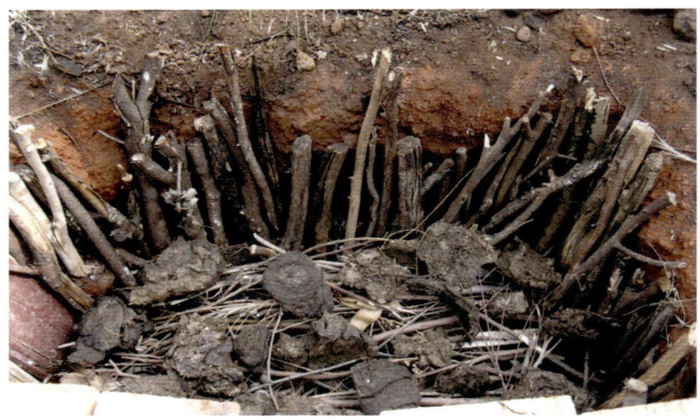

Pit is lined with thick sticks and layered with sticks leaves, cow dung, and seaweed. Oyster shells are also added when available. *Photo: Author*

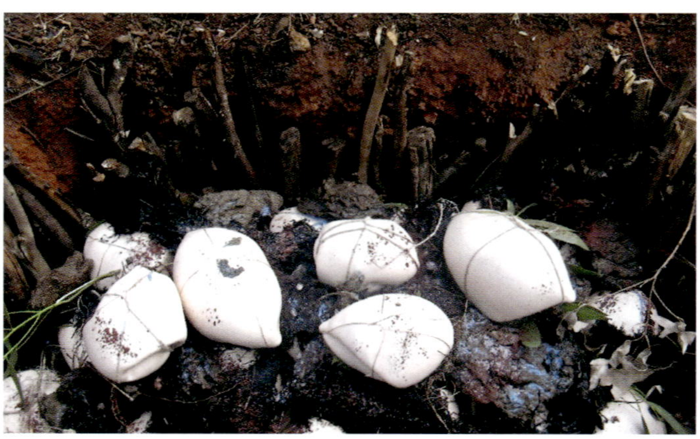

A layer of sculptures is placed in the pit, covered with more cow pats and seaweed, and sprinkled with salt and copper carbonate. A second layer of sculptures is sometimes added and covered with more organic materials. *Photo: Author*

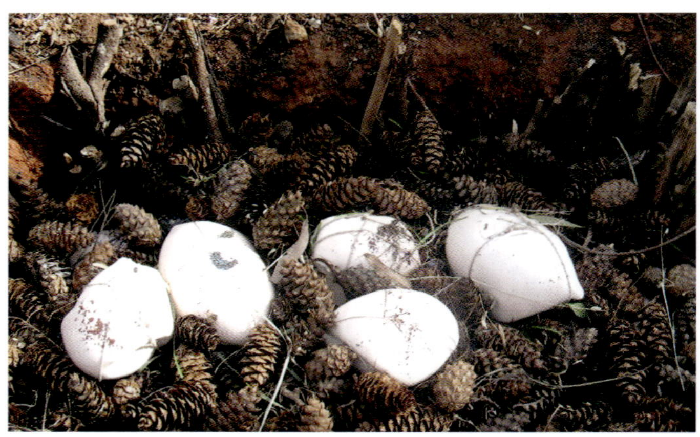

Pinecones are added and finally a layer of branches and wood. *Photo: Author*

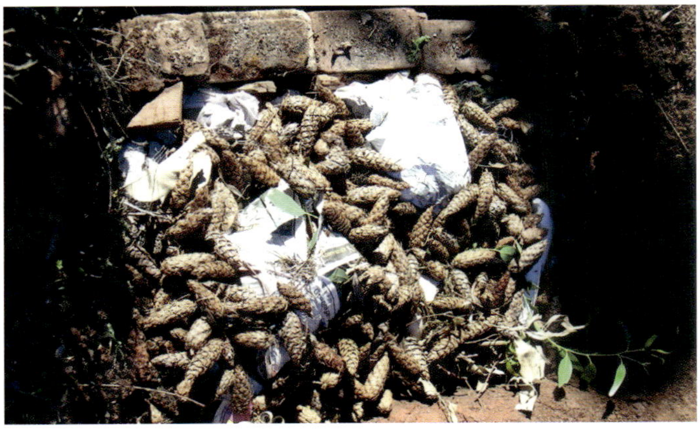

Sometimes works are wrapped in a newspaper saggar containing organic materials and oxides. *Photo: Author*

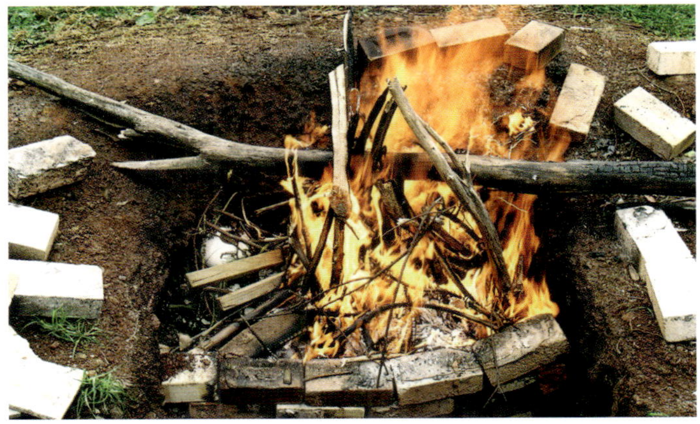

The pit is set alight and stoked with branches and wood for approximately two to three hours. *Photo: Author*

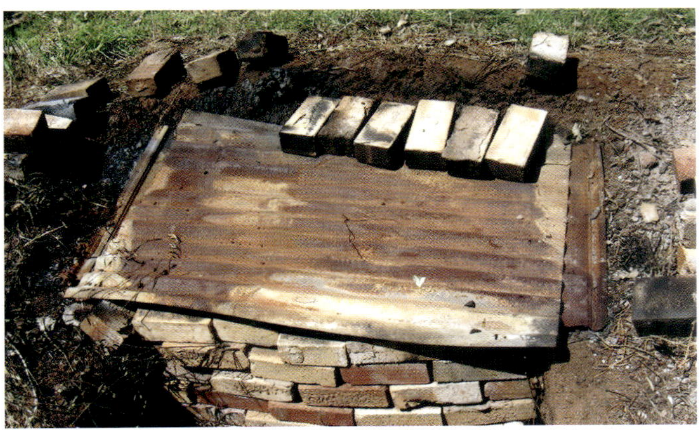

When a thick layer of embers have developed a final layer of wood and branches are added and the pit is covered to allow a slow combustion burn. *Photo: Author*

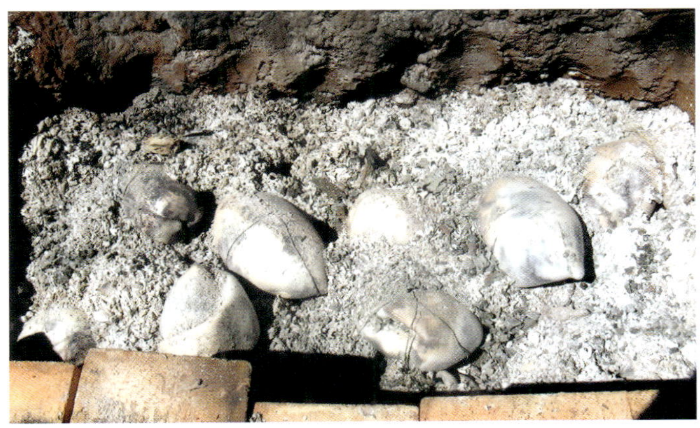

Twenty-four to thirty-six hours later the cover is removed and the pit is unloaded. *Photo: Author*

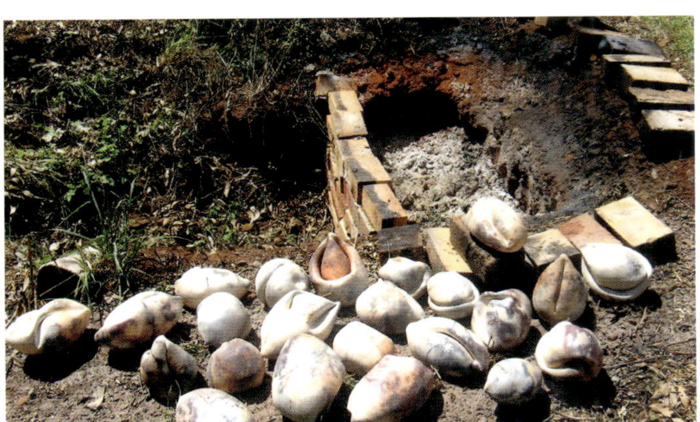

The unloaded pit. *Photo: Author*

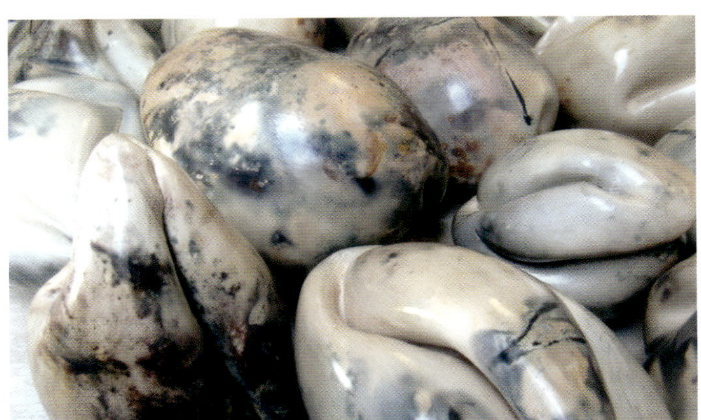

Sculptures are gently cleaned with a damp cloth and/or soft brush and sealed with a wax product. *Photo: Author*

Lauren Bellero

Lauren Bellero fires in a large 10' x 4' x 4' pit. These images illustrate her pit firing process, including three different versions of the loaded pit showing various treatment techniques.

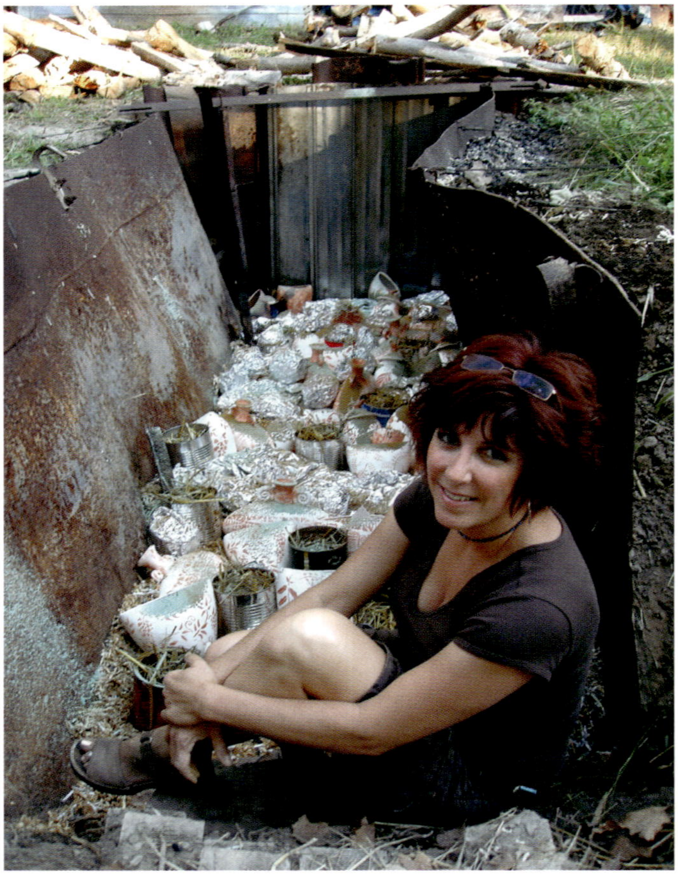

The loaded pit, including the use of aluminium saggars. *Photo: Lauren Bellero*

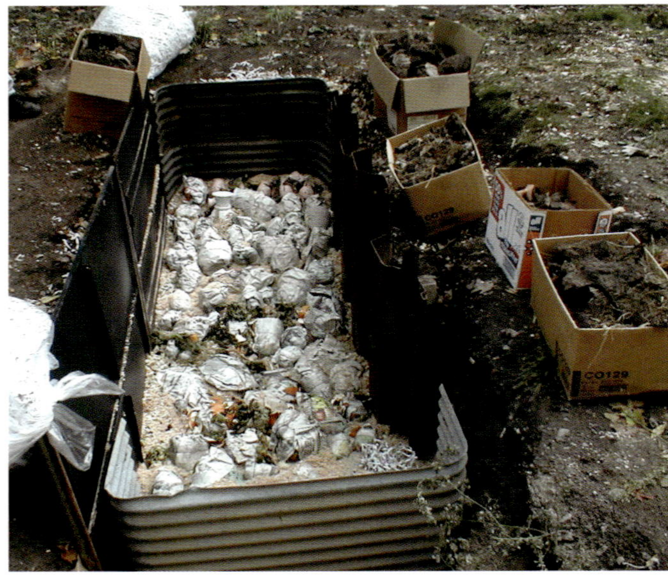

Another version of the loaded pit, including the use of newspaper saggars. *Photo: Lauren Bellero*

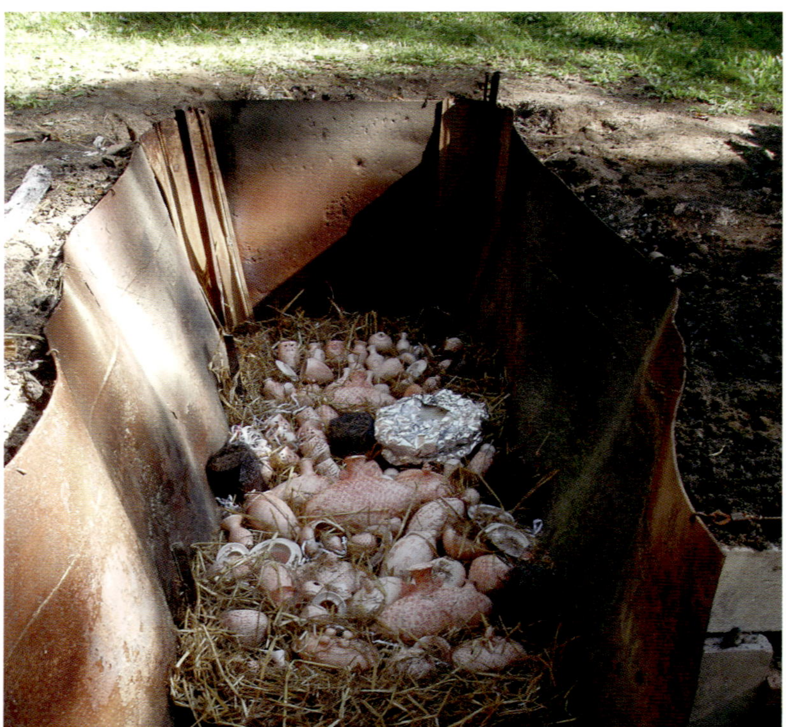

A third version of the loaded pit, including the use of tins to isolate the effects of the firing on pots placed inside the tins. *Photo: Paul Reeser*

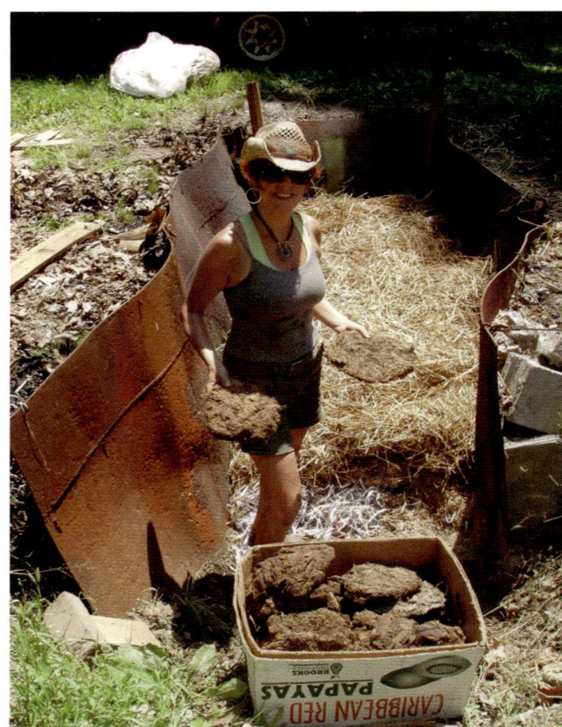

The loaded pit is covered with cow pats and straw. *Photo: Paul Reeser*

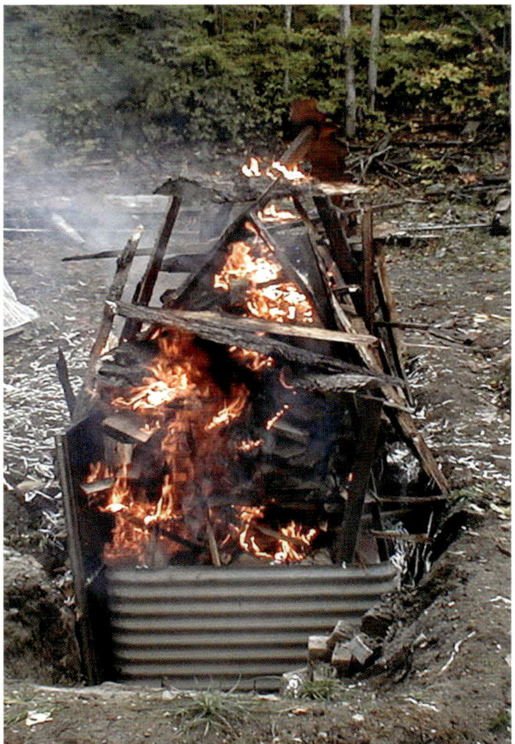
The pit is set alight. *Photo: Lauren Bellero*

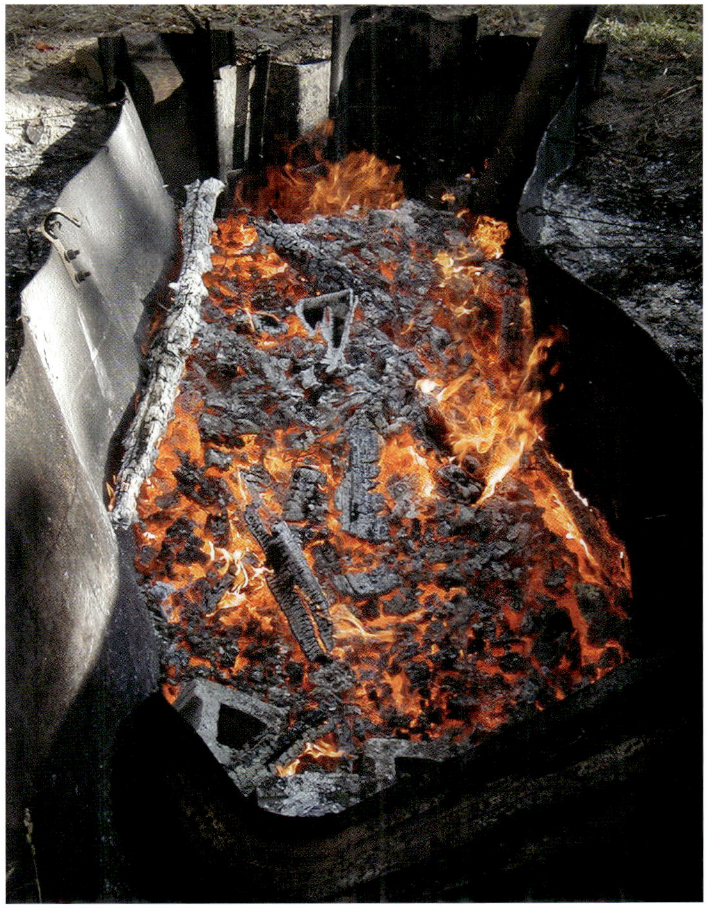
The pit is allowed to burn down to hot embers before being covered for a slow combustion burn. *Photo: Lauren Bellero*

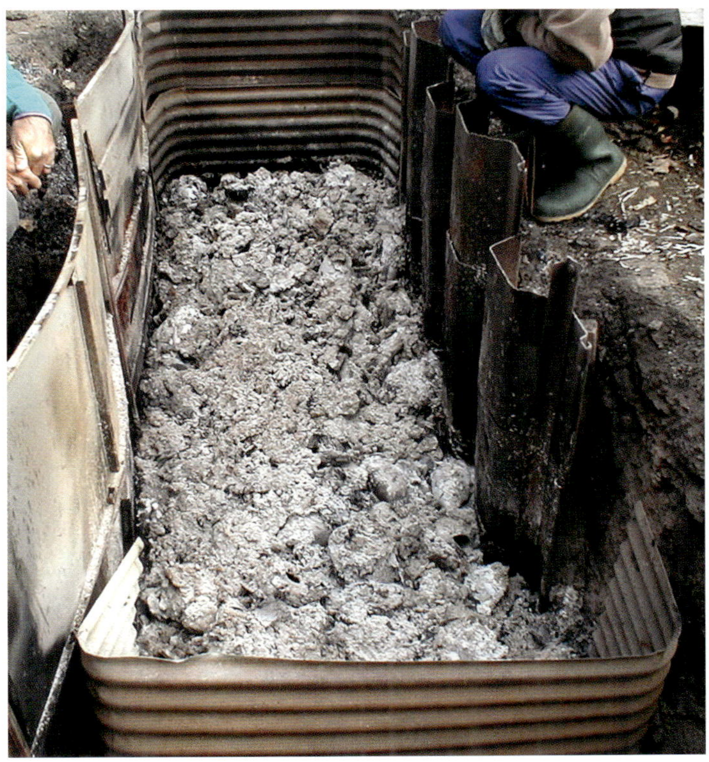
The finished firing ready for unloading. *Photo: Lauren Bellero*

Carol Molly Prier

Carol Molly Prier fires her vessels in a shallow pit on the beach. Each firing pit is custom dug to suit the amount of pots to be fired, and the environmental impact on the site is minimal. Prier will sometimes refire vessels that may not have developed sufficient surface interest in previous firings, an advantageous aspect of pit firing.

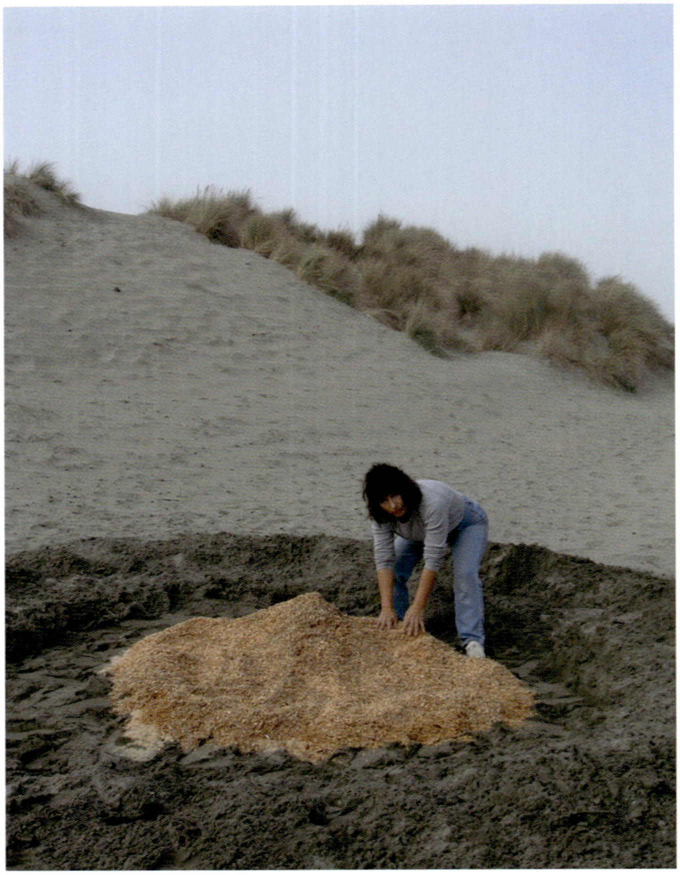

Making a shallow pit on the beach and spreading sawdust. *Photo: Jan Thyer*

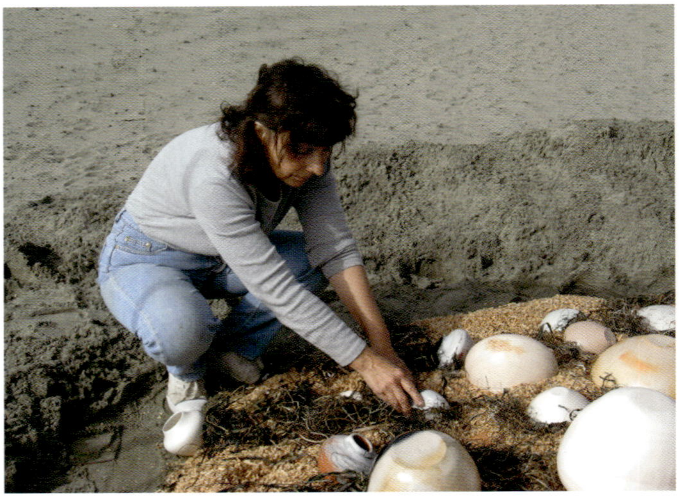

Nestling pots into the sawdust. *Photo: Jan Thyer*

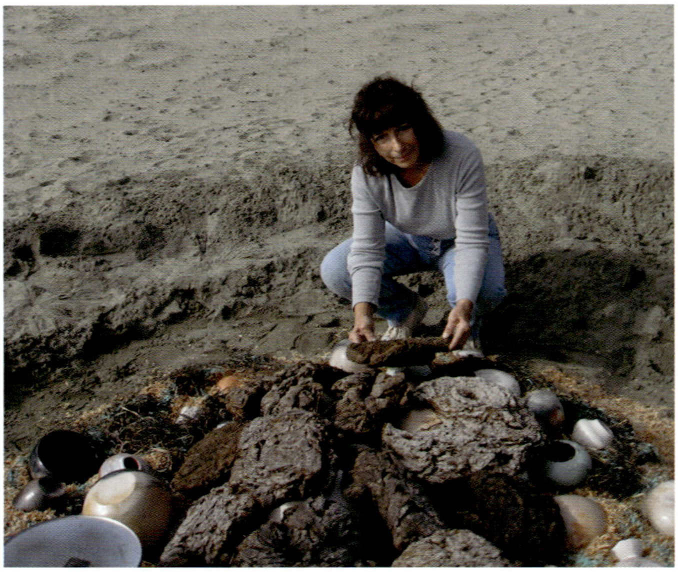

Covering the pit and pots with cow pats. *Photo: Jan Thyer*

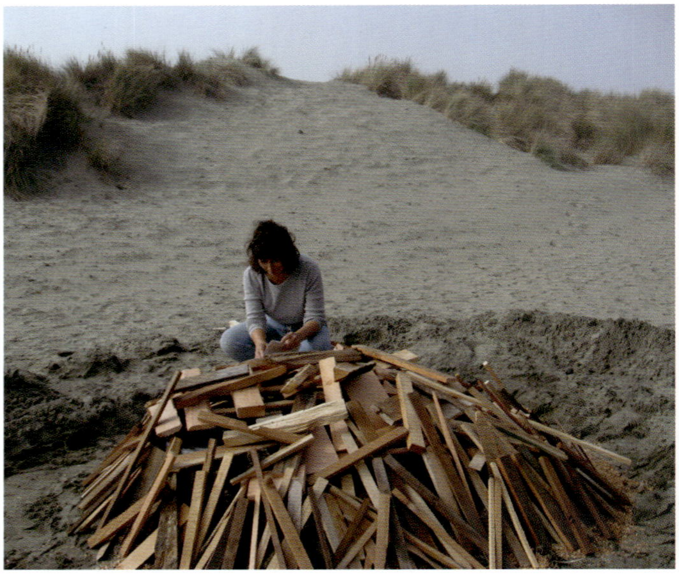

Covering the pit, pots, and pats with wood. *Photo: Jan Thyer*

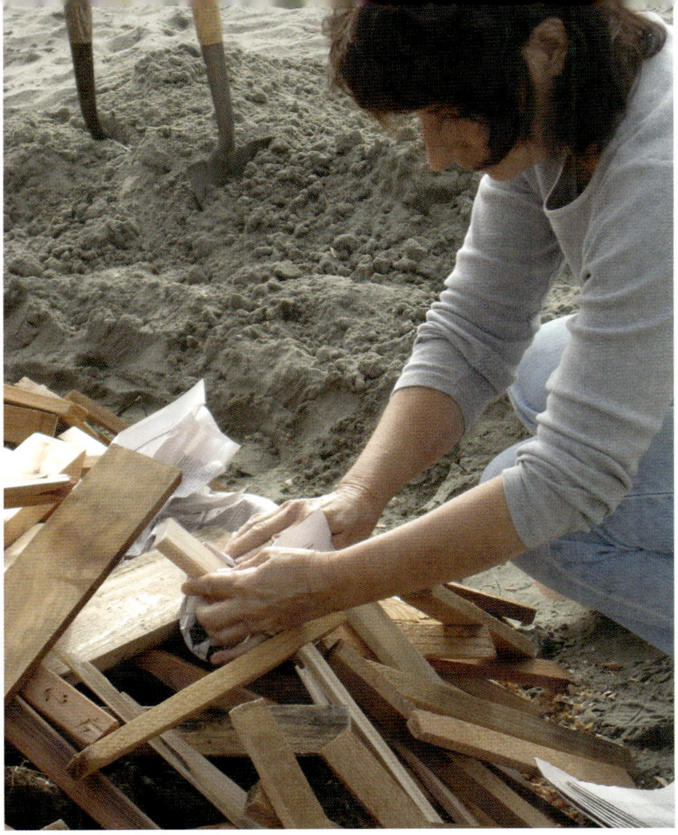

Scrunched up paper is used to light the pit. *Photo: Jan Thyer*

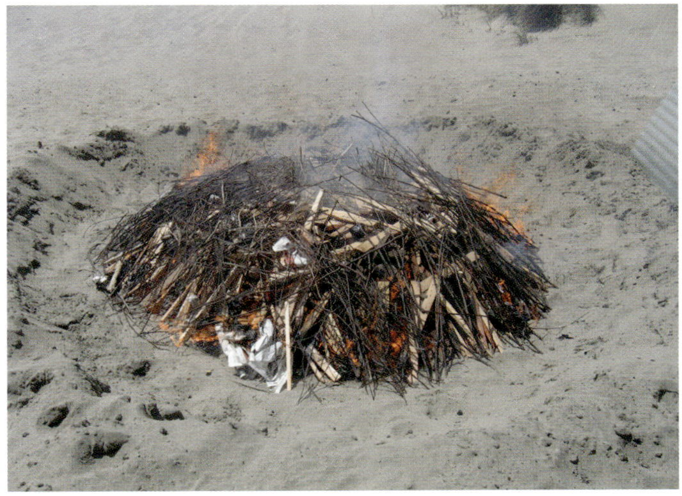

Smaller fuel is added and the pit is set alight. *Photo: Becky Bryant*

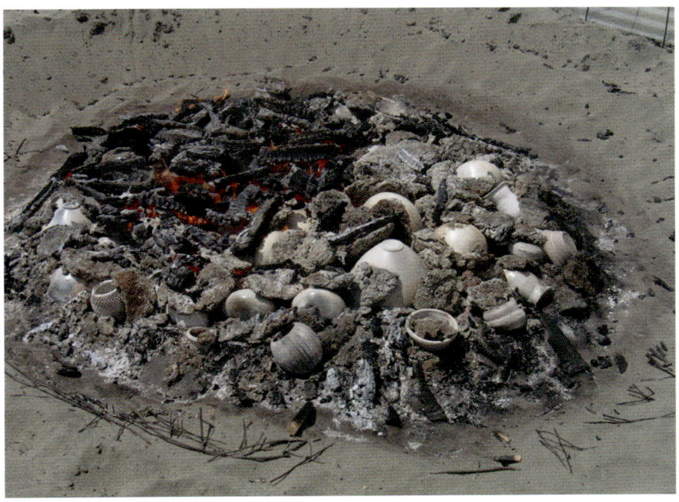

Fuel burns away, leaving embers. *Photo: Becky Bryant*

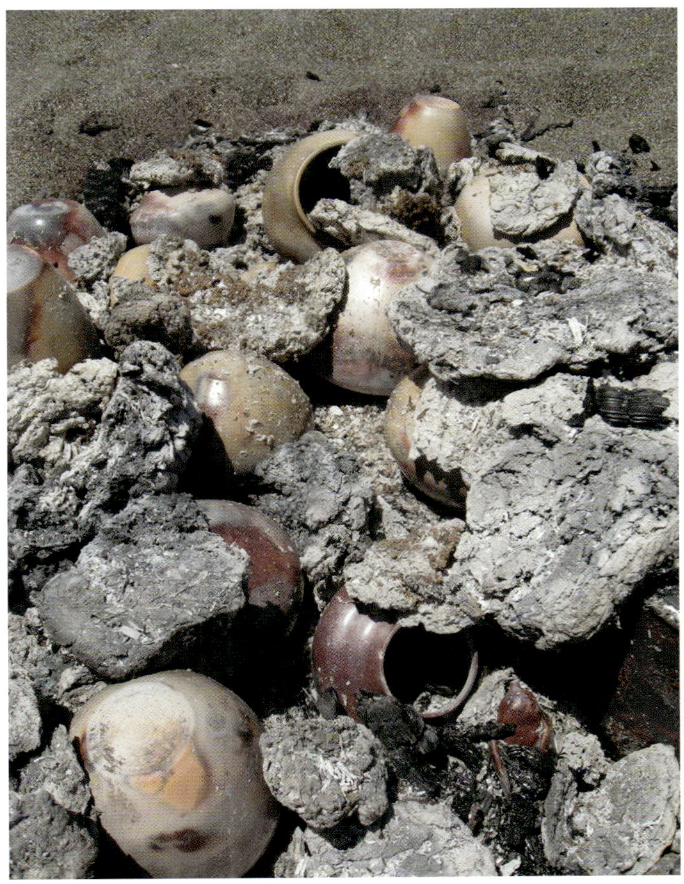

The finished firing. *Photo: Jan Thyer*

David Harder

David Harder pit fires both on the beach and at his property in Arizona. His firings are very successful, which he attributes to the following: a dry pit; plenty of dry hardwood and softwood; well prepared pots—only bisqued to cone 010 and burnished; keeping the pit burning hot till a good layer of coals is established; being patient and gradually covering the pit; and making sure to check the weather forecast!

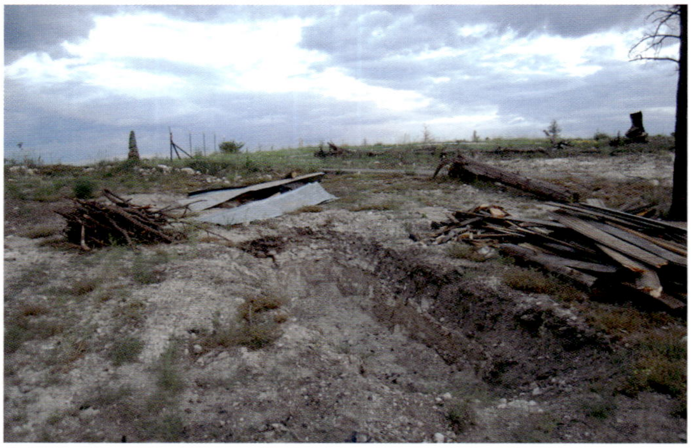

Hole dug with backhoe, 4' wide, 12' long, 4' deep.
Photo: David Harder

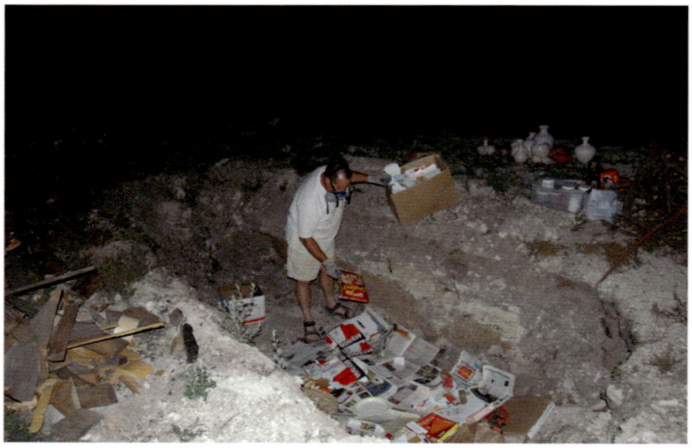

Preparing pit with several layers of newspaper.
Photo: L. Martin

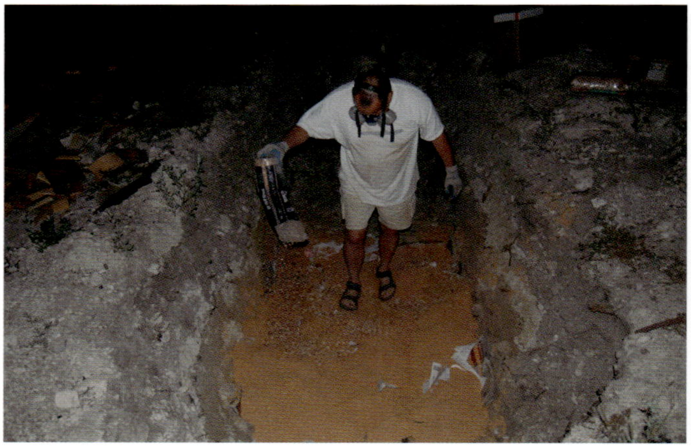

Adding a layer of Juniper sawdust about 4" thick and adding several bags of Cedar chips (store bought at pet shop). A face mask is worn for lung protection. *Photo: L. Martin*

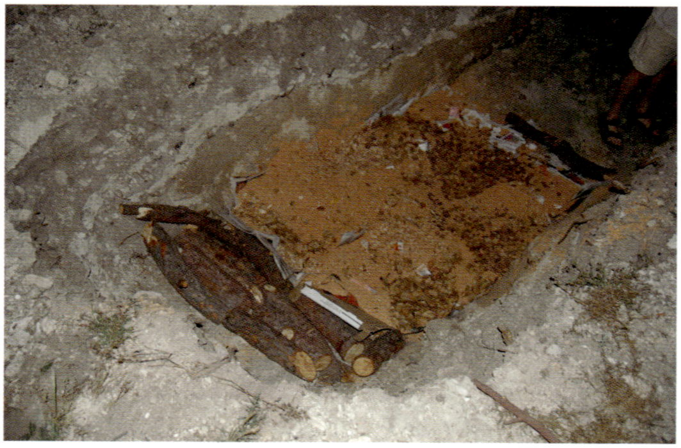

Stacked wood on each end of the sawdust.
Photo: David Harder

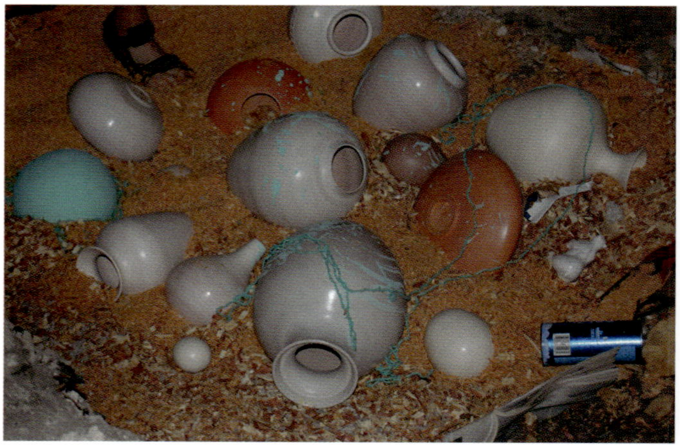

Pots are nestled into the sawdust and string soaked in copper carbonate and shredded copper scour pads are added (Caution—face mask required for chemicals).
Photo: David Harder

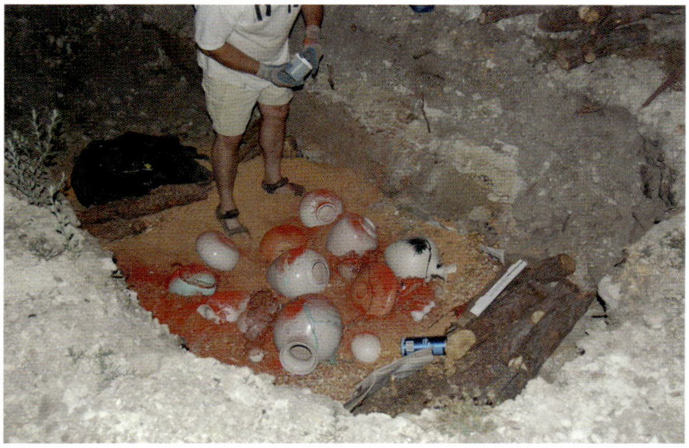

Applying copper, iron, cobalt carbonates and copper; iron and cobalt oxides sprinkled over the pots. *Photo: L. Martin*

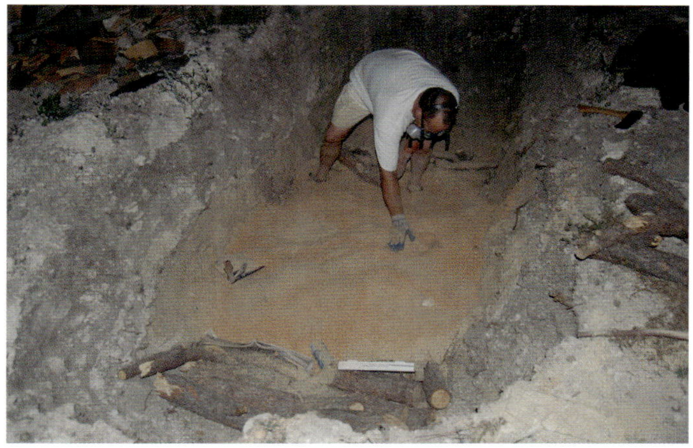

Gently covering pots with juniper sawdust, so as not to disturb the chemicals applied. Ensure the sawdust covers pots by at least 3–4". *Photo: L. Martin*

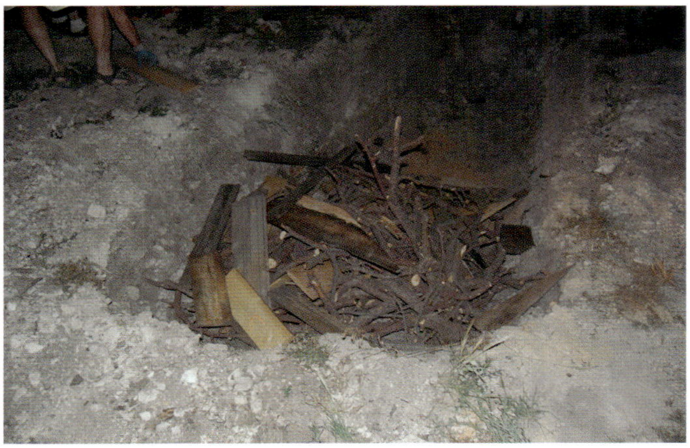

Wood is a mix of pine, juniper, & cedar. *Photo: David Harder*

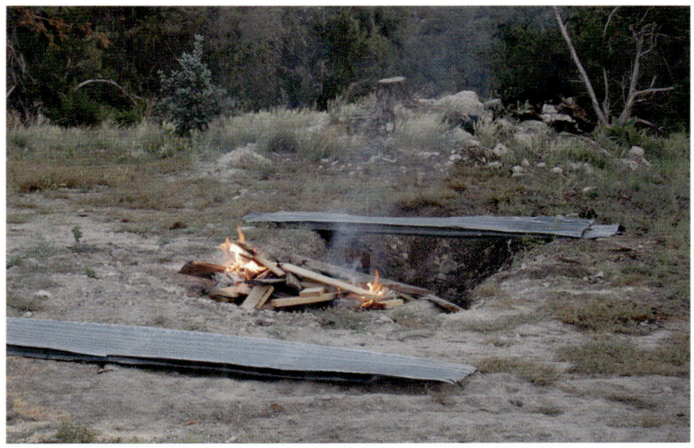

Due to the restriction on burn time, the pit fire is started at 4 A.M. the next day. *Photo: David Harder*

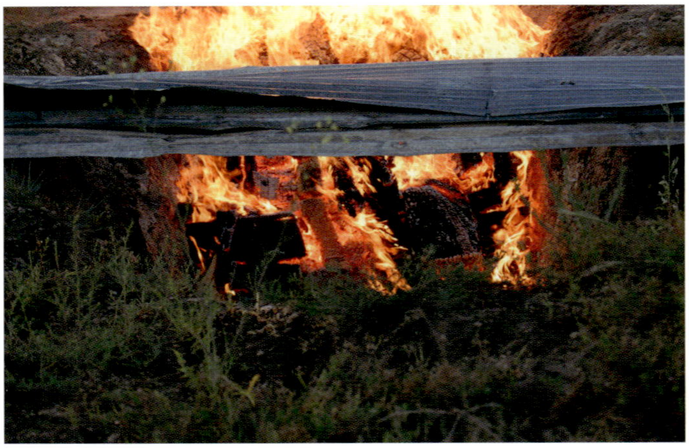

Wood is beginning to burn nice and hot. The flames are monitored, watching for when they die down, the wood turning to coals. *Photo: David Harder*

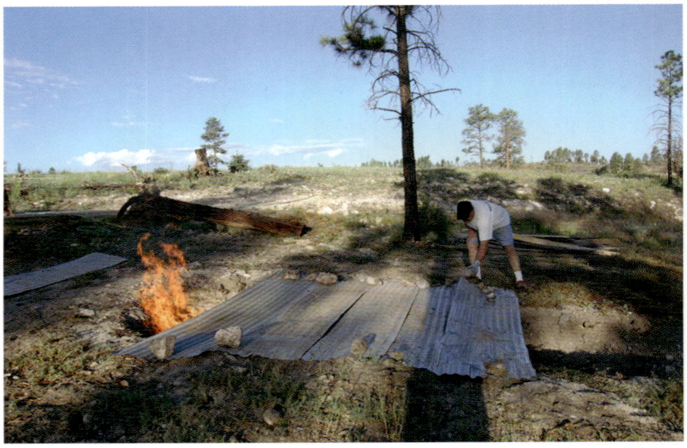

Once the flames have died down and coals remain, the pit is gradually covered with tin. Occasional gusts of wind will blow through, so the tin panels are weighed down with heavy rocks along the edge. *Photo: David Harder*

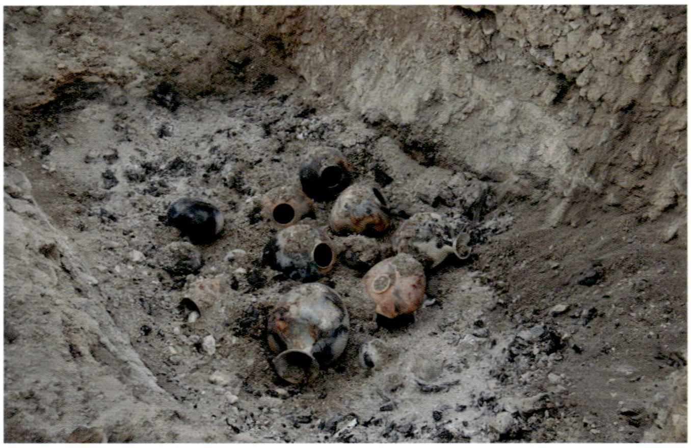

When the pit is uncovered the following day, the ash has settled down, fully exposing the pots, which are still very hot. *Photo: David Harder*

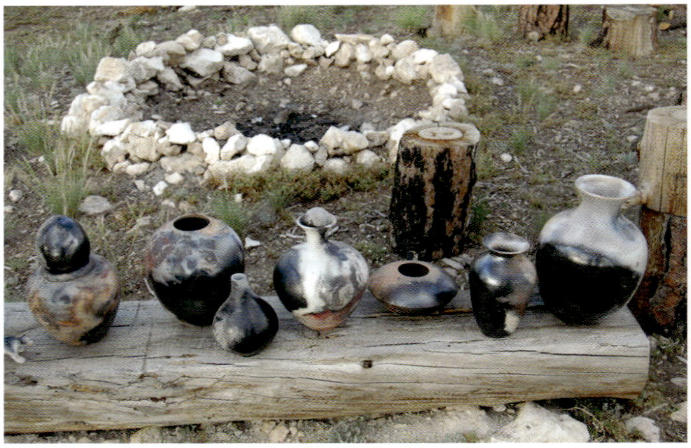

All the pieces from the pit firing, after rinsing with water and set out to dry. *Photo: David Harder*

Up In Smoke Pottery

Up In Smoke Pottery uses a custom blended stoneware clay body that is smoothed with metal ribs while turning on the wheel, coated with terra sigillatta, and lightly burnished. The pots are then bisque fired prior to pit firing.

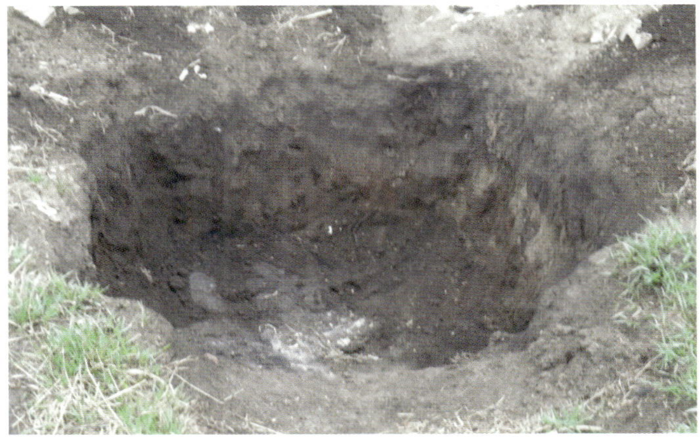

The fire pit: approximately 30 inches deep and 5 feet across. *Photo: Chris Dykstra*

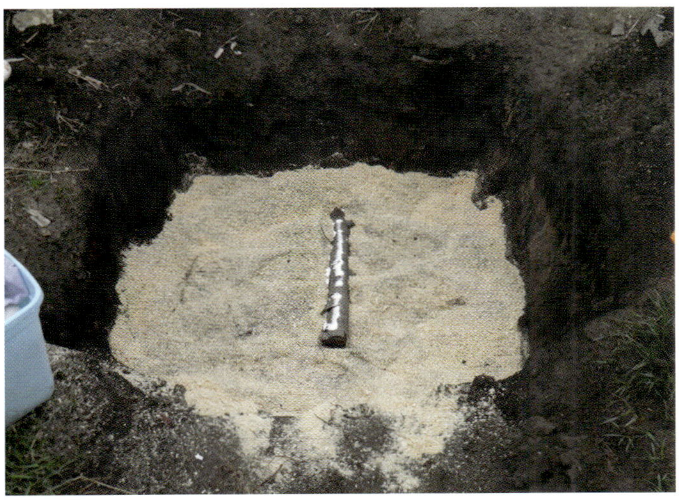

The pit is filled with sawdust and organic materials such as dried leaves, dried grass, wood, metal shavings, salt, and minerals. This photo shows sawdust with the other items mixed in to vary their activation time. There is a log in the center with holes filled with salt. *Photo: Chris Dykstra*

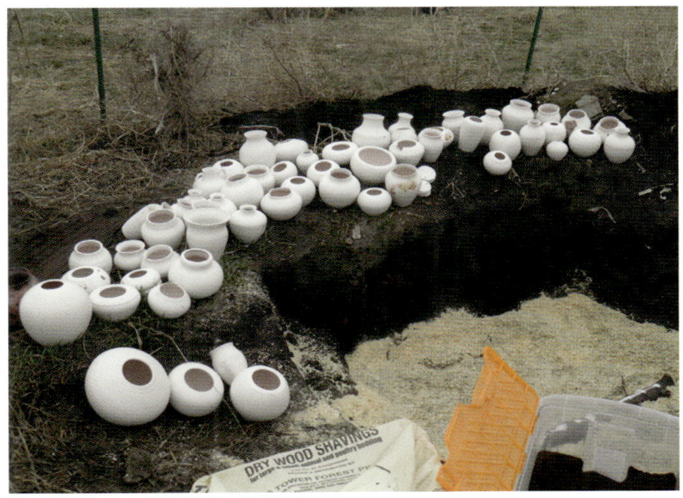

Pots lined up next to pit ready for loading. *Photo: Chris Dykstra*

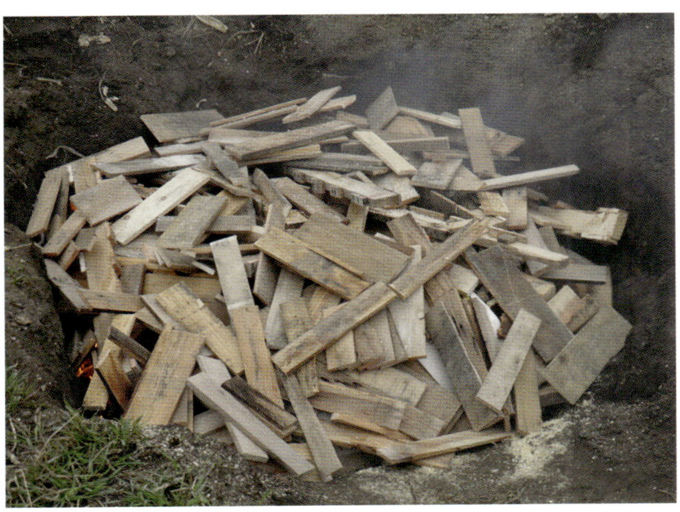

The pieces are buried under paper and wood, taking care not to damage the vessels below. The fire is lit using paper scattered through the wood pile. This image illustrates how much wood is used during each firing. *Photo: Chris Dykstra*

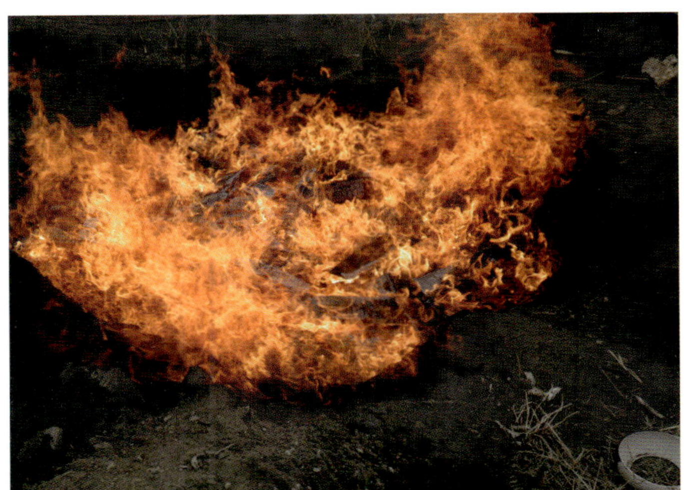

The wood burns very fast and becomes very hot, reaching temperatures of 1400–1800 degrees Fahrenheit. *Photo: Chris Dykstra*

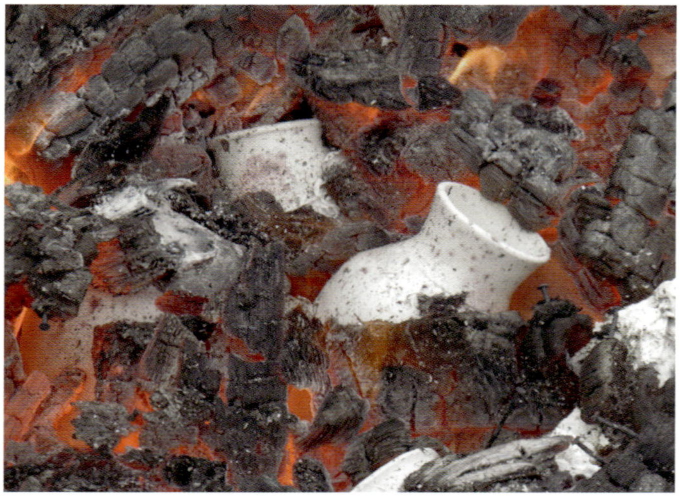

As the fire burns down to a coal bed, pieces begin to appear within the coals. Each piece will appear white until it has cooled enough to show their color. *Photo: Chris Dykstra*

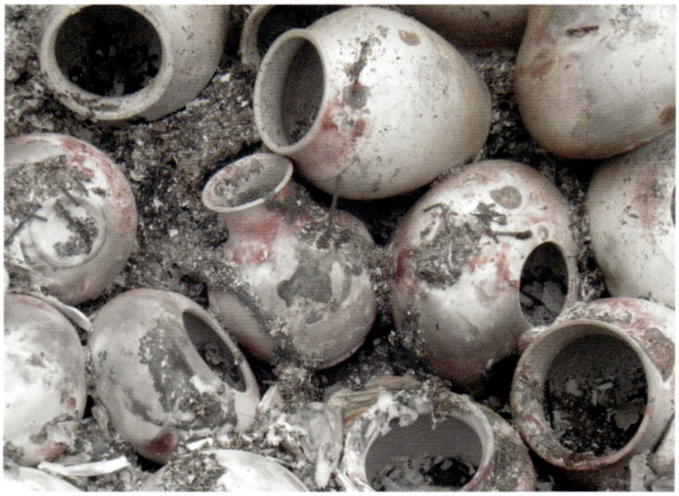

When the fire is completely out, all the pieces will be exposed and have a light covering of ash and soot. Flashes of color can be seen on the pieces in the front of this pit. This photo was taken twenty-four hours after the pit was ignited. *Photo: Chris Dykstra*

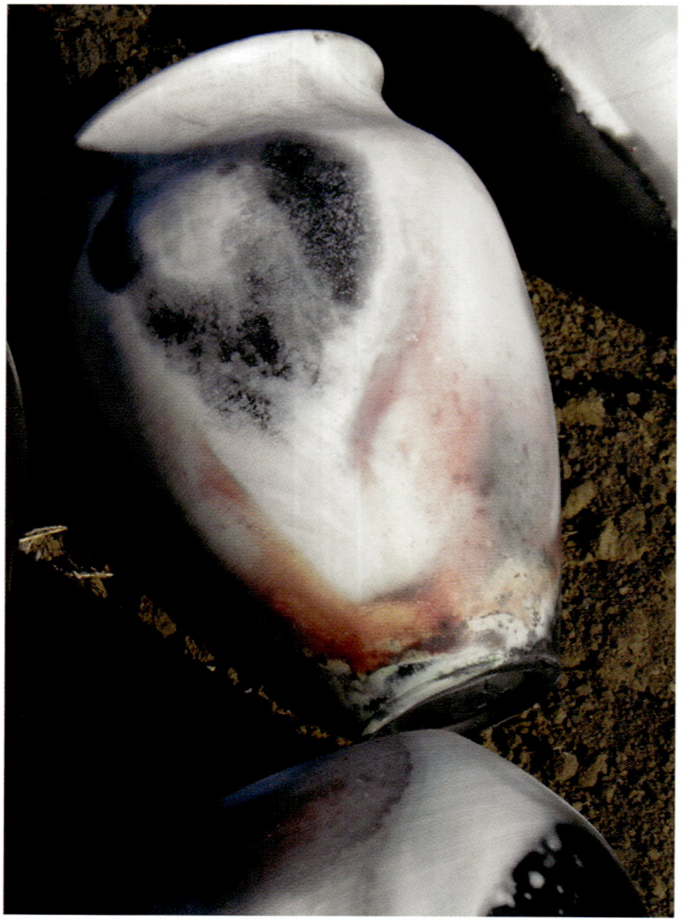

Close up of a piece fresh out of the pit, showing the classic black marking where the piece was nestled in the sawdust, as well as the more subtle vapor marks from where the colored fumes moved around the piece. *Photo: Chris Dykstra*

Maeva Collins

A high sheen on her pots is very important to Maeva Collins, as it gives her pieces visual depth rather than the pattern from the fire appearing to sit on the surface. Collins fires in a brick lined pit.

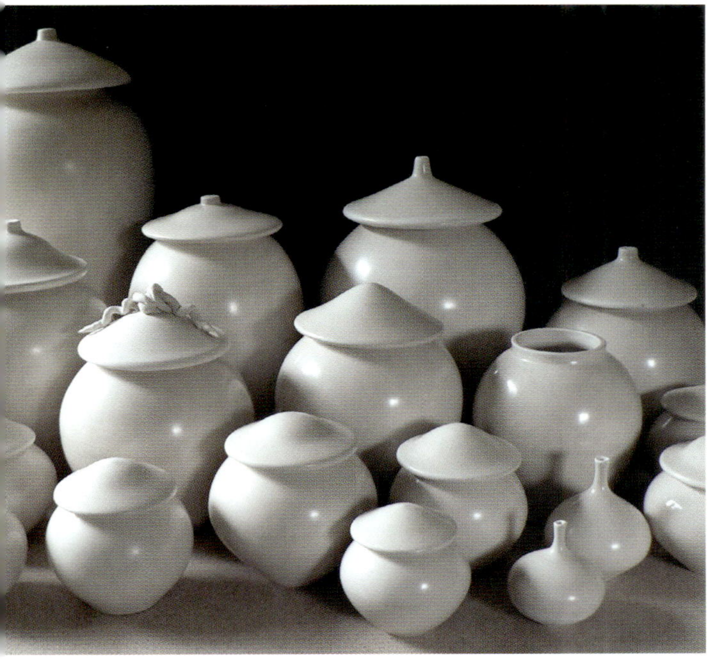

After the second burnishing, the pots are bisque fired in an electric kiln to cone 012. If they were fired any higher, it would cause the pots to loose their shine. *Photo: Maeva Collins*

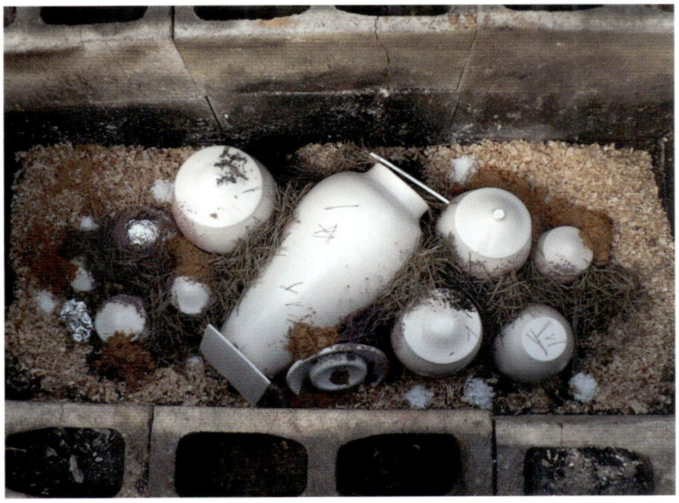

A three inch layer of shavings is put on the bottom of the pit and then the pots are nestled into the shavings. Each piece is layered with combustibles such as dog food, pine needles, fine sawdust, coffee, or anything else that is available. Sometimes copper carbonate is added for possible color. When the layering is complete, at least 10–12 inches of shavings are added to cover the work. *Photo: Maeva Collins*

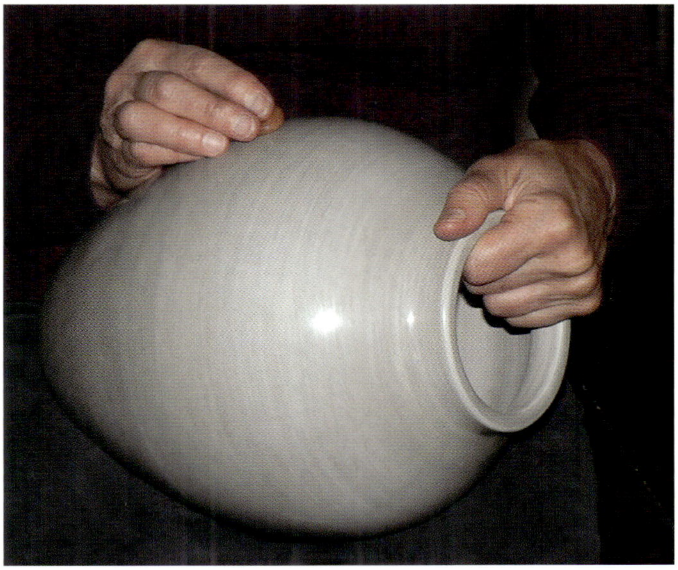

Each vessel is burnished twice, once when it is leather-hard and then again when it is bone dry. This picture shows the second burnishing. Various shapes and sizes of agates and quartz stones are used during this process. The double burnishing results in a very shiny surface. *Photo: Maeva Collins*

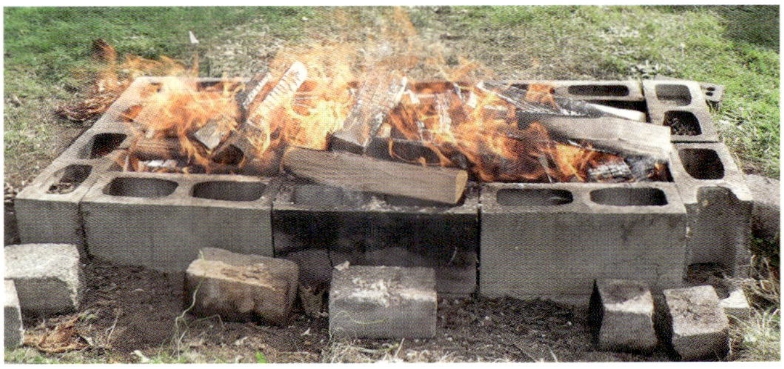

Paper and kindling is added on top of the sawdust and the fire is started. Alder and maple firewood are usually used, because they are available. More wood is added until large amounts of hot embers are generated. The metal lid is dropped and the fire is allowed to burn down for at least two days. *Photo: Maeva Collins*

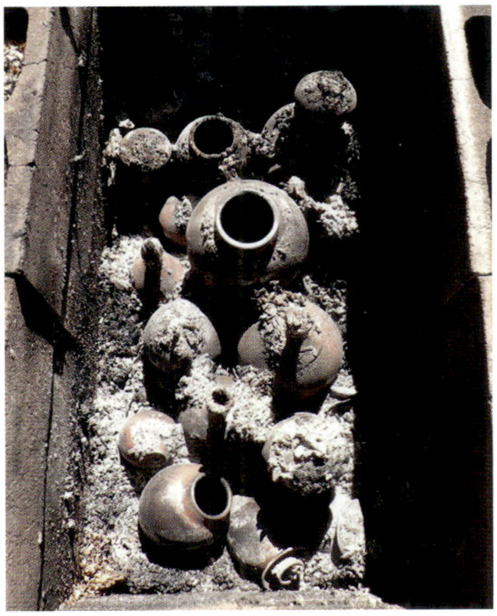

When cool, the pit is uncovered and the pots are washed. When completely dry, they are finished with a coat of paste wax and polished to protect the surface. *Photo: Maeva Collins*

HEALTH AND SAFETY ISSUES

The very nature of pit firing means that the artist is coming into direct contact with fire and smoke, therefore a degree of caution should always be undertaken. Non-flammable, reasonably tight fitting clothing should be worn, and hair tied back. The site of the pit should be chosen carefully, avoiding nearby fences or trees that could catch fire. Safety equipment, such as a hose or fire extinguisher, should also be readily available. When handling the various oxides used in the pit, I wear thick gardening gloves and a ventilator mask, as some oxides can be absorbed through the skin or inhaled. Once the pit is set alight, the mask is kept on while stoking the pit, as some substances, such as salt, can become toxic upon reaction with the flames. Most countries have fire regulations at different times of the year, for example in Australia there are always fire restrictions in the summer months, so special permits need to be sorted through with the fire authorities. If in doubt, check with your local council.

CHAPTER FOUR
ARTIST PROFILES

The art of pottery is, of all the arts, the one that fuses together indestructible unity, earth and heaven, and matter and spirit (Koyama 1975, 5).

ANNE ARMSTRONG
(CANADA)

Anne Armstrong enrolled for her first pottery lesson in 1974, and has been potting ever since. A full time professional potter for thirty years, she established her studio, Brickworks Pottery, in 1981. At that time, she made functional stoneware pottery. In 2003, Armstrong realized she was burnt out from the constant demand for functional wares and decided to make ceramics for herself. Armed only with some books and online guides, she enlisted a friend with a backhoe to dig a hole in her yard, and fired her first pit. The results were mixed, but one little vase exhibited a strong slash of burgundy running diagonally up the pot, bordered by white and black. Since then she has not looked back from pit firing and other experimental firing techniques. Armstrong adds a range of organic ingredients to her pit, including steel wool, copper wire, copper sulfate, copper carbonate, epsom salts, Miracle Gro®, dog and cat kibble, and occasionally horsehair. Armstrong is represented in several public and private collections and had her work *Reliquary* published in *500 Raku* (Lark Books) in 2011.

Of her work, Anne Armstrong says:

> I am a maker of objects. I find excitement in the making and satisfaction in the completion of my creations.
>
> Pleasing proportions, confident forms, and clean unbroken curves are all important aspects of my process. Then, casting predictability aside, I trust my works to the vagaries of fire and wind. Primitive firing methods remove my ability to control the final outcome.
>
> That people find beauty in my work is a welcome by-product of my need to create.

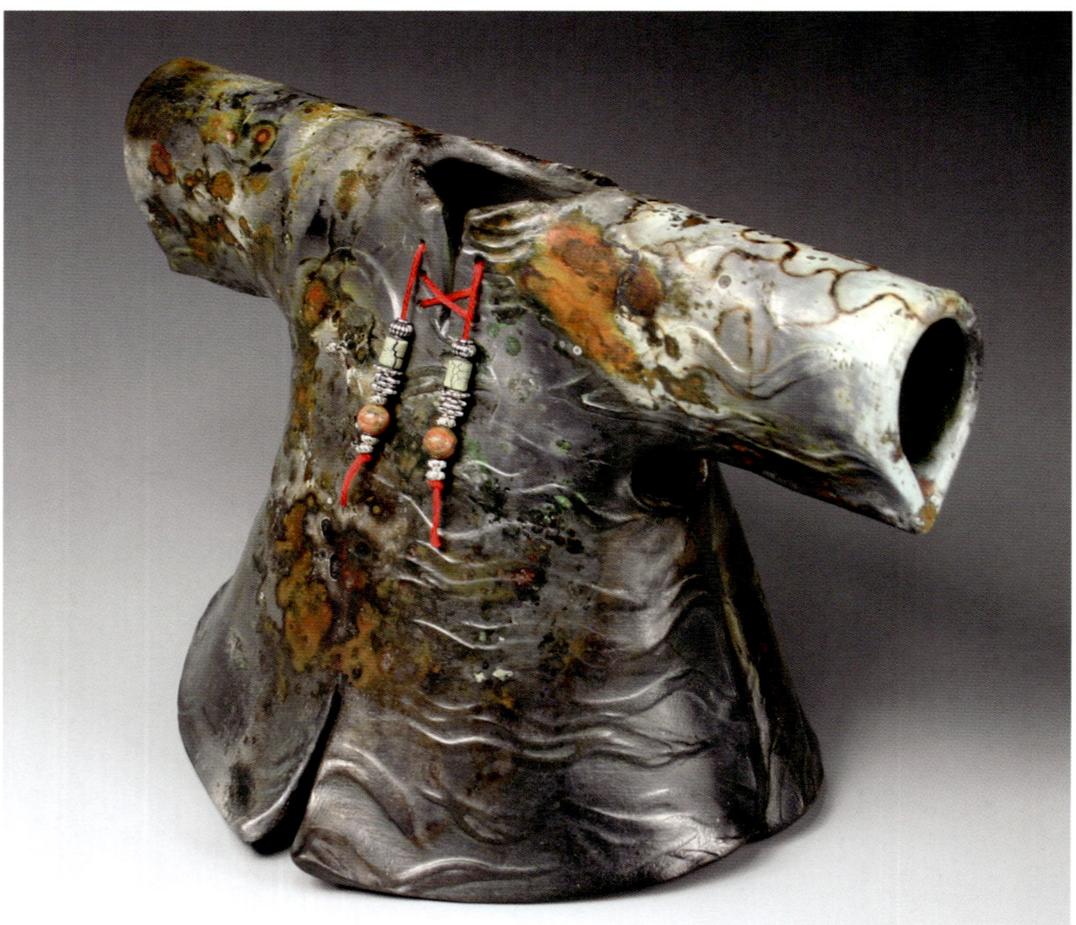

Anne Armstrong. *Kimono.* Handbuilt white stoneware, terra sigillata, 5 ½" h x 8" w. *Photo: Anne Armstrong*

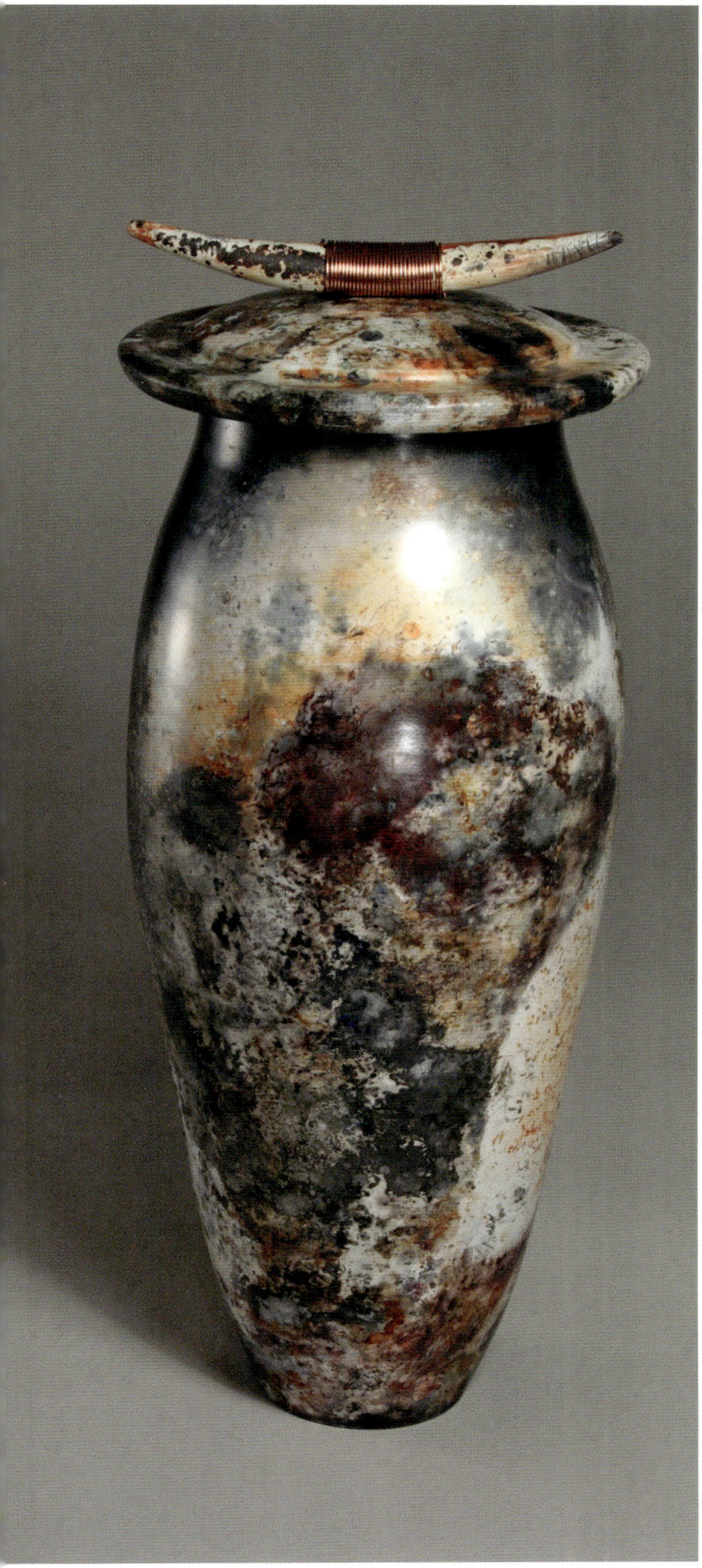

Anne Armstrong. *Rage On.* Wheel thrown porcelain, terra sigillata, 14" h x 5" d. *Photo: Anne Armstrong*

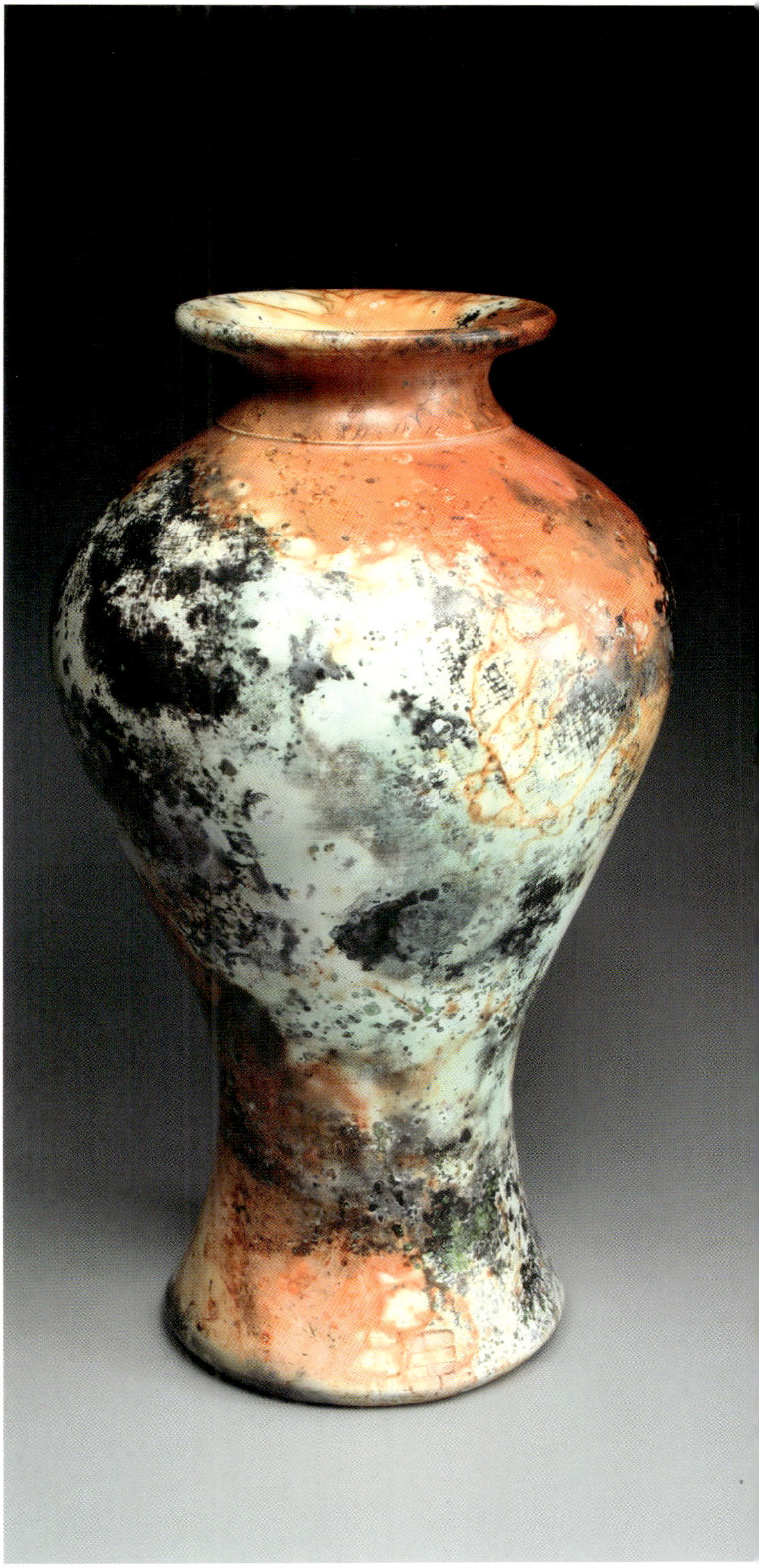

Anne Armstrong. *Journey.* Wheel thrown porcelain, terra sigillata, 10" h x 4 ½" d. *Photo: Anne Armstrong*

BENNETT BEAN
(US)

"Some boys outgrow their love of fire. I don't seem to have managed."

Bennett Bean is best known as a ceramic artist (for his treatment of vessels post firing), but works in a variety of media, including stone, precious metals, wool and silk weaving, paper, parchment, and painting. He was formally trained in fine art and has been making art, prolifically, since 1960. For the last twenty-five years he has been full time in his studio. Bean's work is represented by numerous galleries as well as in major American museum collections, including the Whitney Museum of American Art and the White House Art Collection.

Of his art practice Bennett Bean says:

> To understand an object, I want to connect with it, to live with it, to have it around me. I'll buy an object if I can afford it, but some things don't exist anywhere but in my head. Those I have to make. In making, I learn what's there. The things I make influence what I buy and the things I buy influence what I make. From this process objects accumulate. Then comes the problem of "putting a thing in the world." How do you present a pot, a painting, a piece of sculpture? You need some place to put it. So I work on the house. I don't make any distinctions between making things, cooking, gardening, and building houses. Elements from the garden appear in paintings and the surface obsession of the pots appears in the house as consciousness of each decision about material and finish. Each cross-pollinates. Curiosity about how to express identity resulted in having my DNA analyzed. That image then surfaced in collages and then again in rugs. The paintings and the pots have both contributed their imagery to the rugs. It's a dance where ideas are applied in different ways depending on the medium.

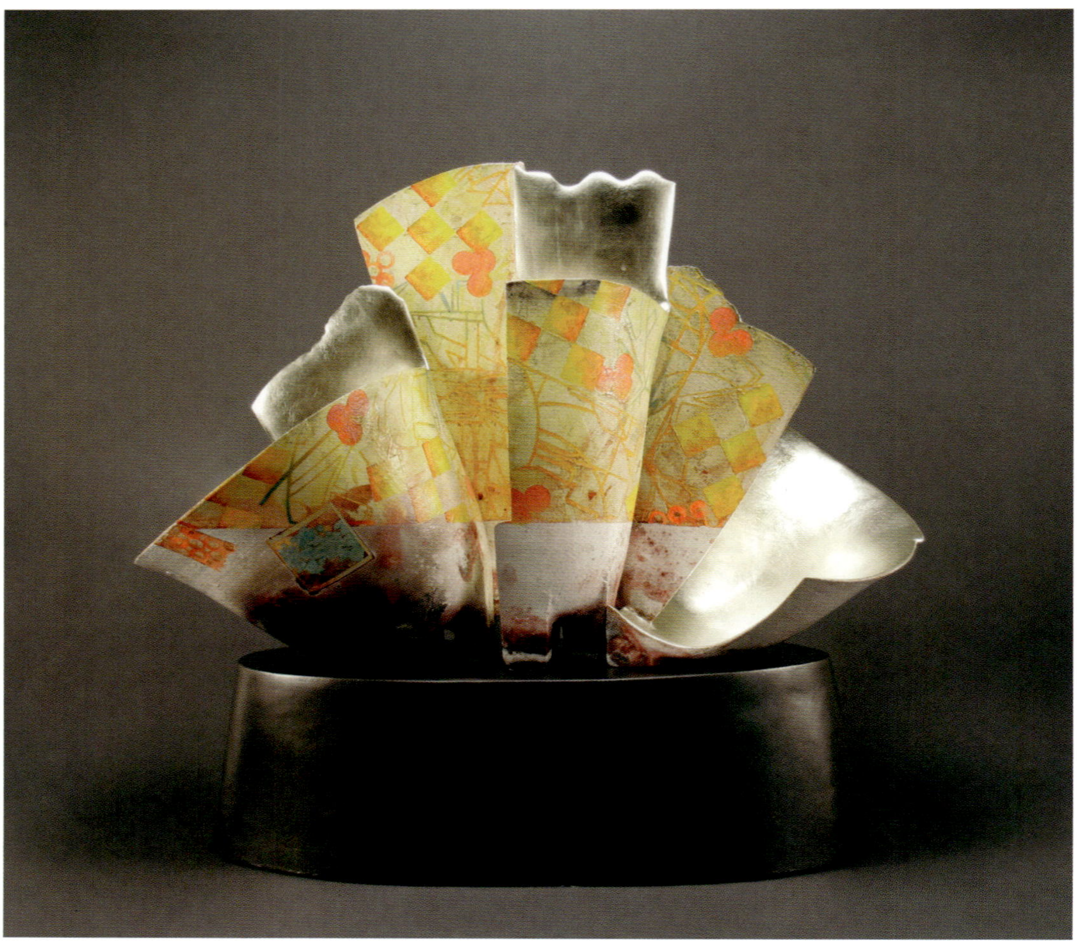

Bennett Bean. *Master #1617.* Triple on base. Thrown and altered, terra sigillata, pit fired, painted and gilded earthenware. 31 ½" w x 29 ¼" h x 14 ¾" d. *Photo: Bennett Bean*

Bennett Bean. *Master #1618.* Four part on base. Thrown and altered, terra sigillata, pit fired, painted and gilded earthenware. 29 ¾" w x 31 ½" h x 14.5" d. *Photo: Bennett Bean*

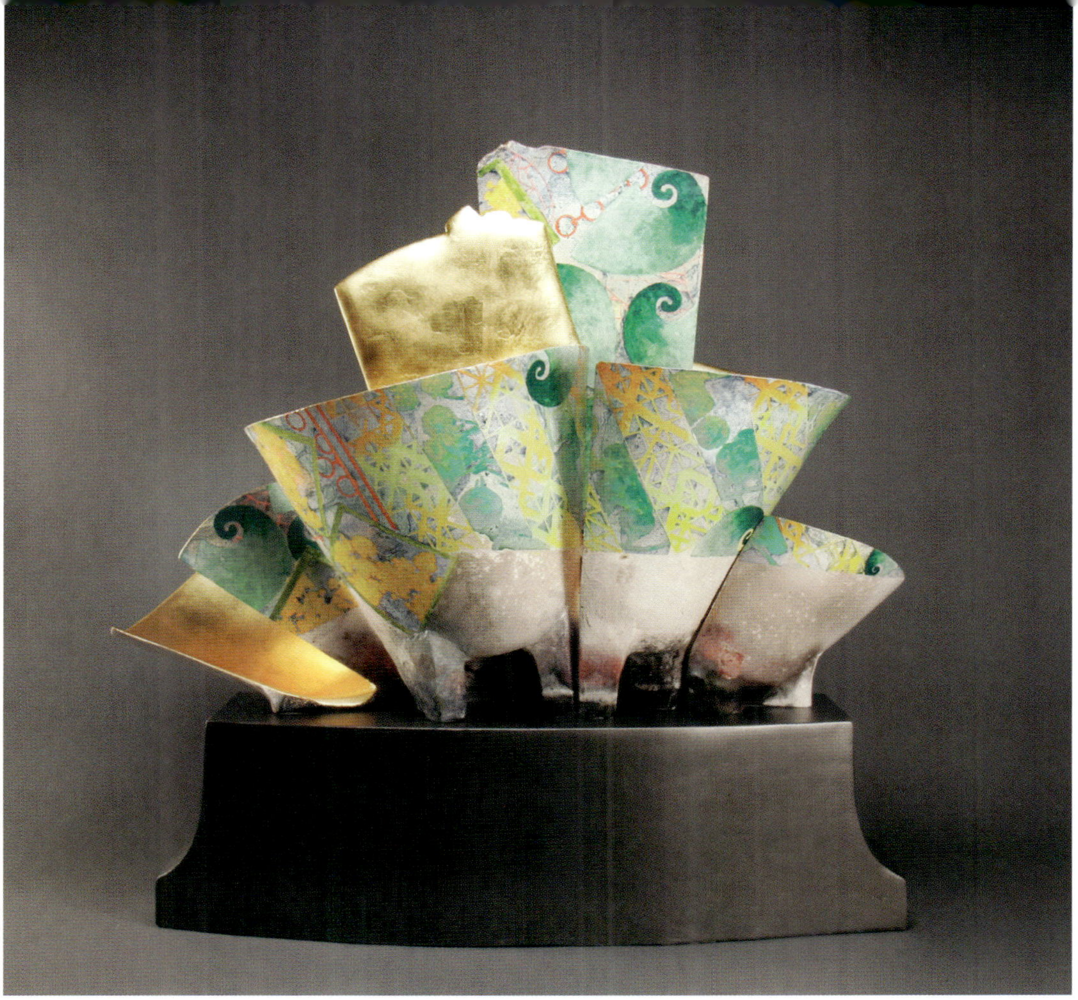

Bennett Bean. *Master #1619.* Four part on base. Thrown and altered, terra sigillata, pit fired, painted and gilded earthenware. 32" w x 28" h x 13" d. *Photo: Bennett Bean*

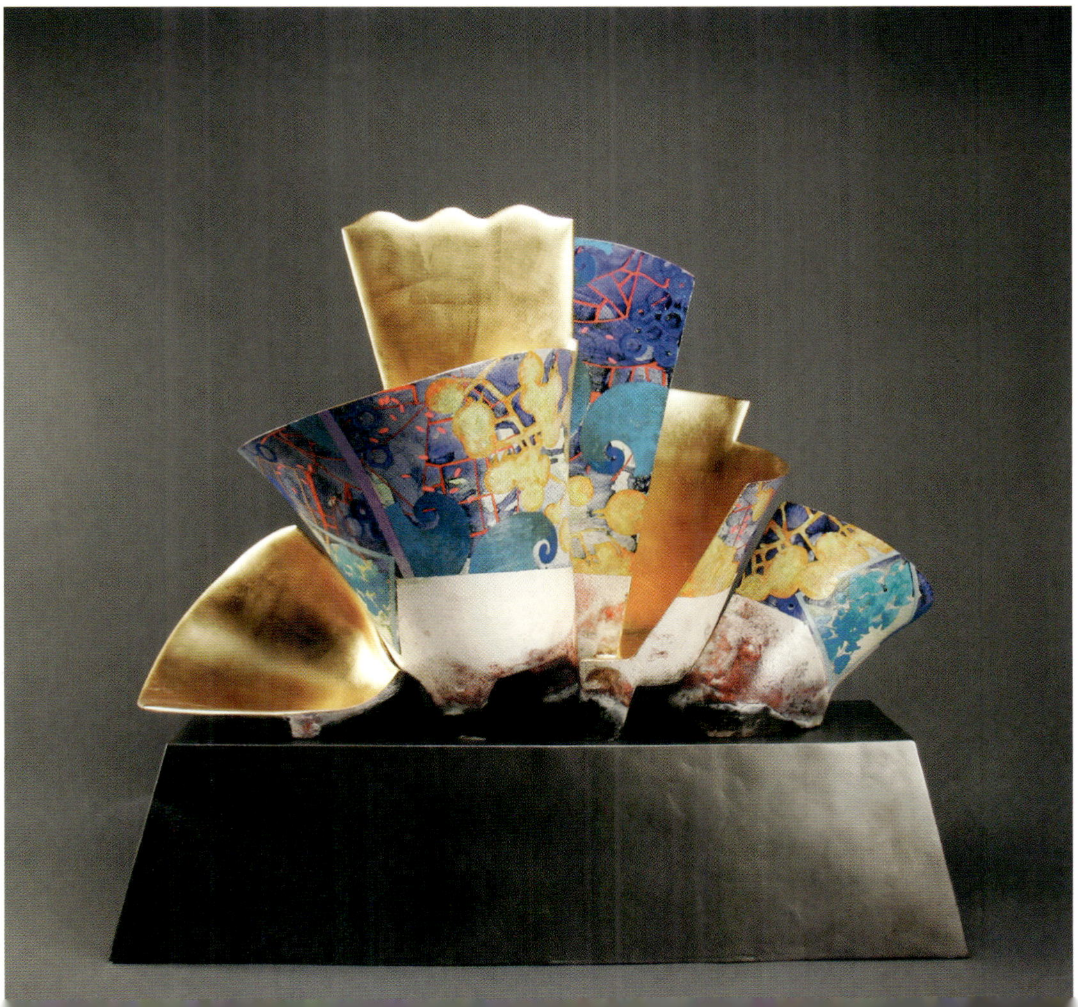

LAUREN BELLERO

(US)

In her own words: Hmmm ... my "rationale" for selecting pit firing ...

Well, the short and sweet answer is that sometimes glazing for me is almost painful! No, not in the physical sense, it's just that it is labor intensive to "think" about it, to make and "test" it, to do it (multiple sprayed layers), and to have it come out the way I'd like it to!

I glaze fire in an electric kiln. Electric kilns are easier to fire, but a little more difficult to get satisfactory results. The reason I actually get to pit fire is my husband loves to fish. We have a cabin in Canada (no television, electric, etc.—thank God for the flush toilet!). In order to get me to enjoy going there more often, he had a big pit dug for me so that I would get to have some fun while he was fishing!

Lauren Bellero fires all her pit fired ceramics in a large 10' x 4' x 4' pit.

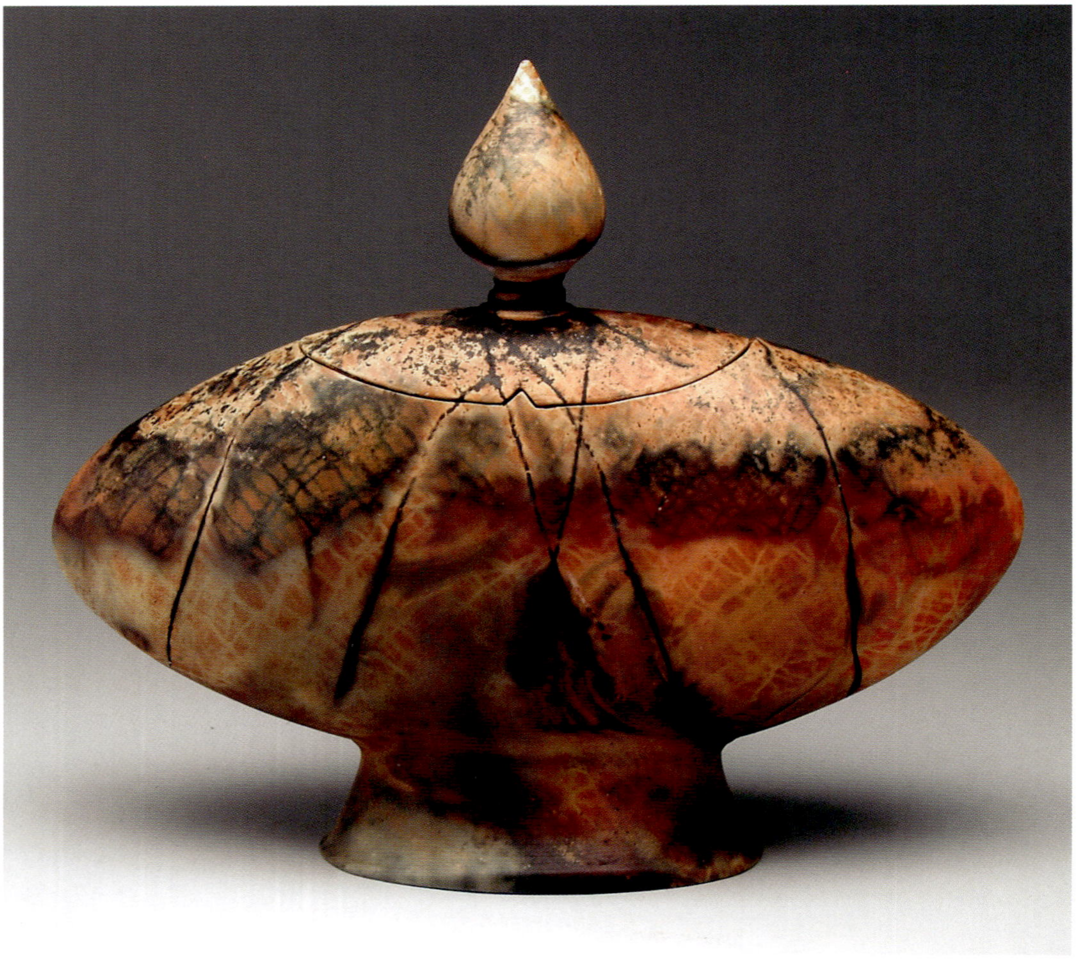

Lauren Bellero. *Pit Fired Urn.* Slab built with hollow, thrown knob, thrown and altered foot, cut-in lid, terra sigillata surface with red terra sigillata pattern, soft burnished. 14" x 12" x 4". *Photo: Lauren Bellero*

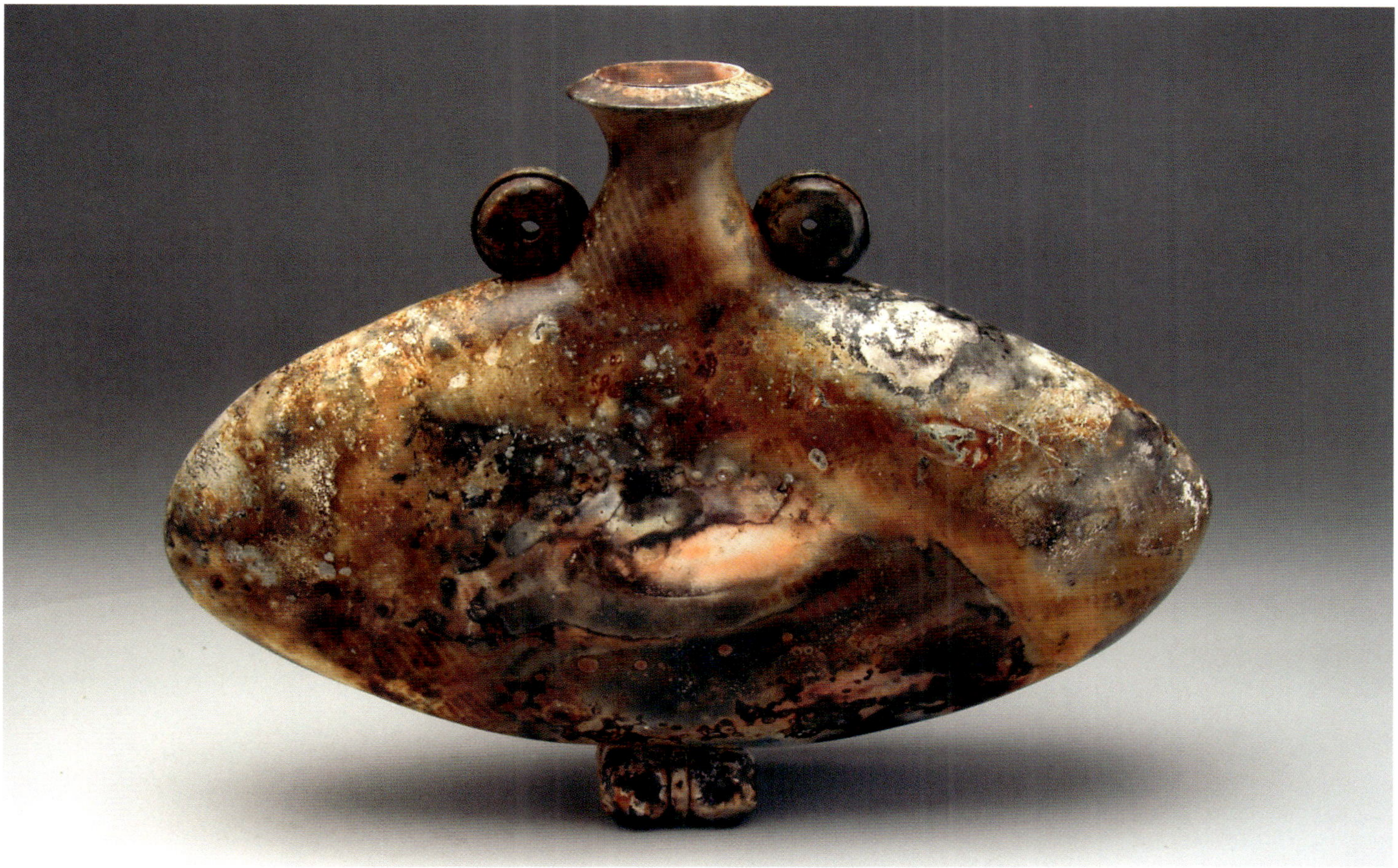

Lauren Bellero. *Pit Fired Vessel.* Slab built with thrown neck, added feet, and lug decoration, terra sigillata surface with red terra sigillata pattern, soft burnished. 12" x 16" x 3". *Photo: Lauren Bellero*

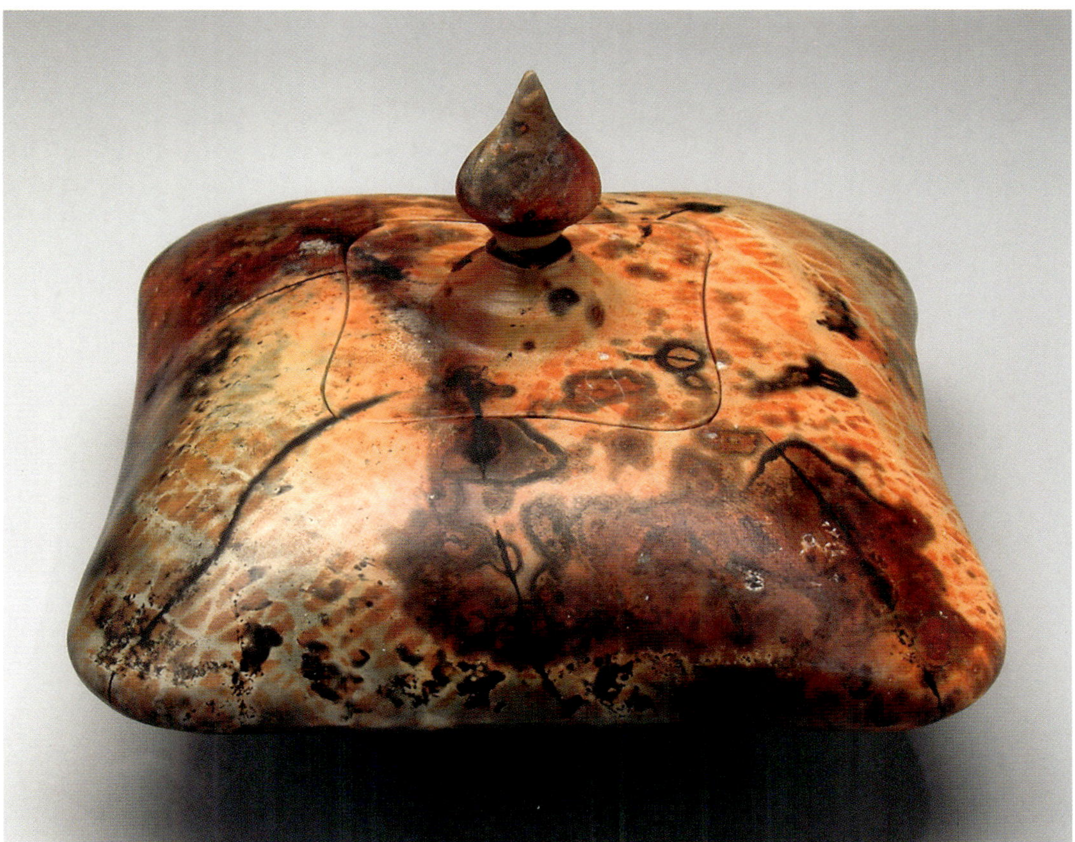

Lauren Bellero. *Pit Fired Pillow.* Slab built with thrown knob, added feet, cut-in lid, terra sigillata surface with red terra sigillata pattern, soft burnished. 8" x 8" x 6". *Photo: Lauren Bellero*

MICHAEL BERKLEY
(US)

Michael Berkley has been teaching and displaying his works for over thirty years. His interest and study in archaeology, and the ancient Sumer culture of the Middle East, and their round clay tablets, began early in his career and continues to influence his work.

The work Berkley produces centers around inner activity; its reach from the root of Self, and the activities that subsequently rise to the surface and leave their marks. Whether from absolutely perfectly thrown porcelain or stoneware clays (during his production days), to his present day work, Berkley continues to explore all aspects of this medium. Berkley's ability to express is shown through his wheelwork, sculpture, and workshops. He is known for being a natural teacher with the ability to reach people at all levels of skill.

Berkley has been in California for over fifteen years and has been teaching at the Walnut Civic Arts Centre for more than twelve. His unique teaching style and techniques for beginning students enables him to teach them how to center three to five pounds of clay in less than fifteen seconds, and he is often asked to teach his "anyone can center" workshop. Berkley teaches beginning, intermediate, and advanced throwing as well as raku and pit firing classes and workshops.

Of his art practice, Michael Berkley says:

> Pit firing has been used as a method of finishing ceramic pieces for thousands of years. Each firing is unique and lends itself to different results.
>
> The artist has control in the early stages of making his wares. When the piece reaches the final stage in this type of firing, one can only direct, through experience, a suggested path for the flame and color to proceed on. This is called "Flame Painting." This medium offers spontaneous surface decoration and is particularly suited to the task of learning how to let go—a typical life lesson for us all.
>
> The pieces are burnished with polished rocks in the leather-hard and bone dry state to achieve a smooth surface or terra sigillata is applied to the surface—the same method used by early artisans.
>
> I choose this type of firing because during the process, at any time, situations can change. This suits me well. I have always been interested in archaeology. Of great interest to me have been the Sumerian Tablets. Writings have been scratched on clay from the beginning of time. Whether it was a crude map or an early inscription, the earth, or clay, was the first to be used and reaches back into our instinctual history.
>
> Some of the earliest writings found on clay were the Sumerian Tablets dated around 4000 BC (and older), which accurately relate the stories of celestial planets that exist in our solar system, complete with their moons, their dominate gaseous or aquatic features, their colors, and their sizes; all of these being accurate. Most of these tablets however had everyday stories: loves, losses, and transactions. The stories give us a window into the past, convey an insight into Now and Then, and are separated only by the construct of time. These writings, or stories, preserve connection with our own deep human history, our own personal history, and leave our "mark" at that moment in time.

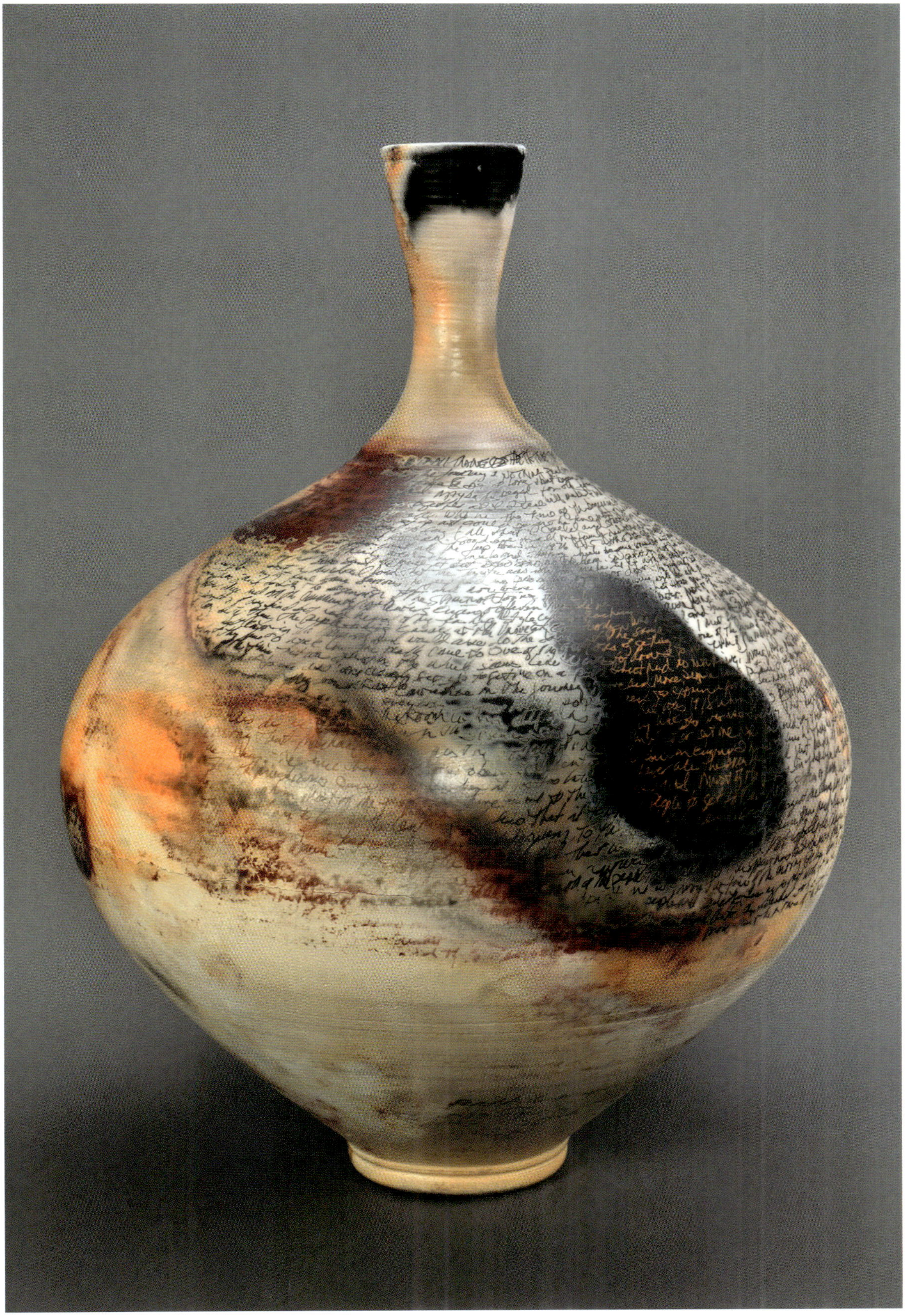

Michael Berkley. *HisStory.* Wheel-thrown and altered with applied carrot orange terra sigillata with silver-white terra overlay. Copper, dung, salt, gold, and copper inscription. Cherry, walnut, pine, oak, and madrone wood with Berkley's specialty mix sawdust. 22" h x 17" w. *Photo: Christine Ferrell*

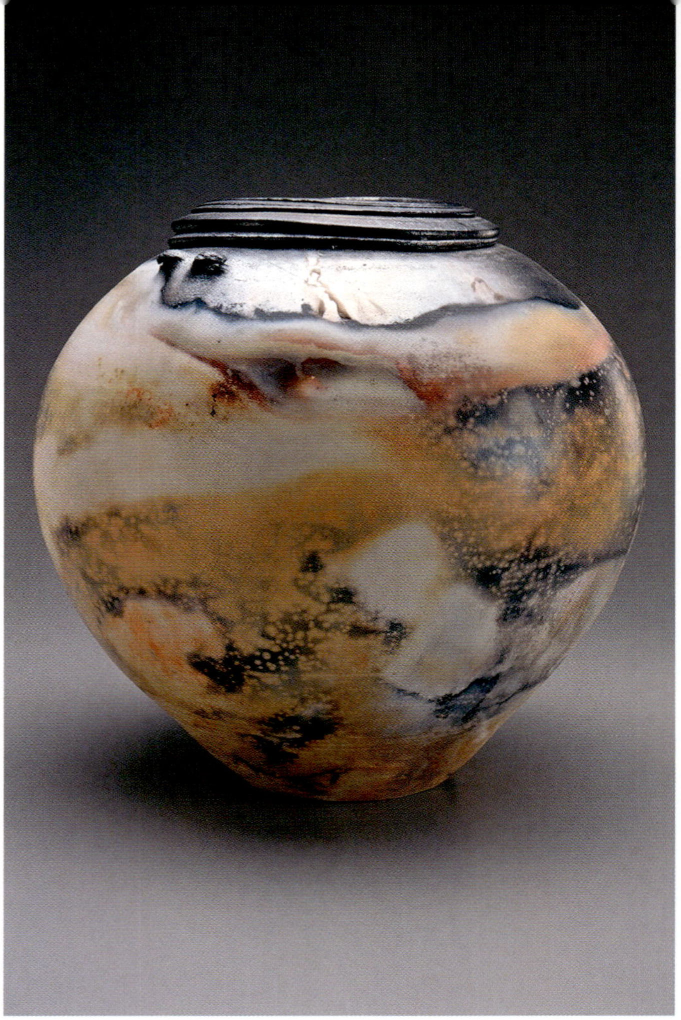

Michael Berkley. *Starry Night*. Wheel-thrown and altered with silver-white terra sigillata and iron mix, copper carbonate, Nick's special dung with holes, sea salt, cherry, walnut, and cedar wood, and Berkley's special mix sawdust. 14" h x 13" w. *Photo: Sibila Savage*

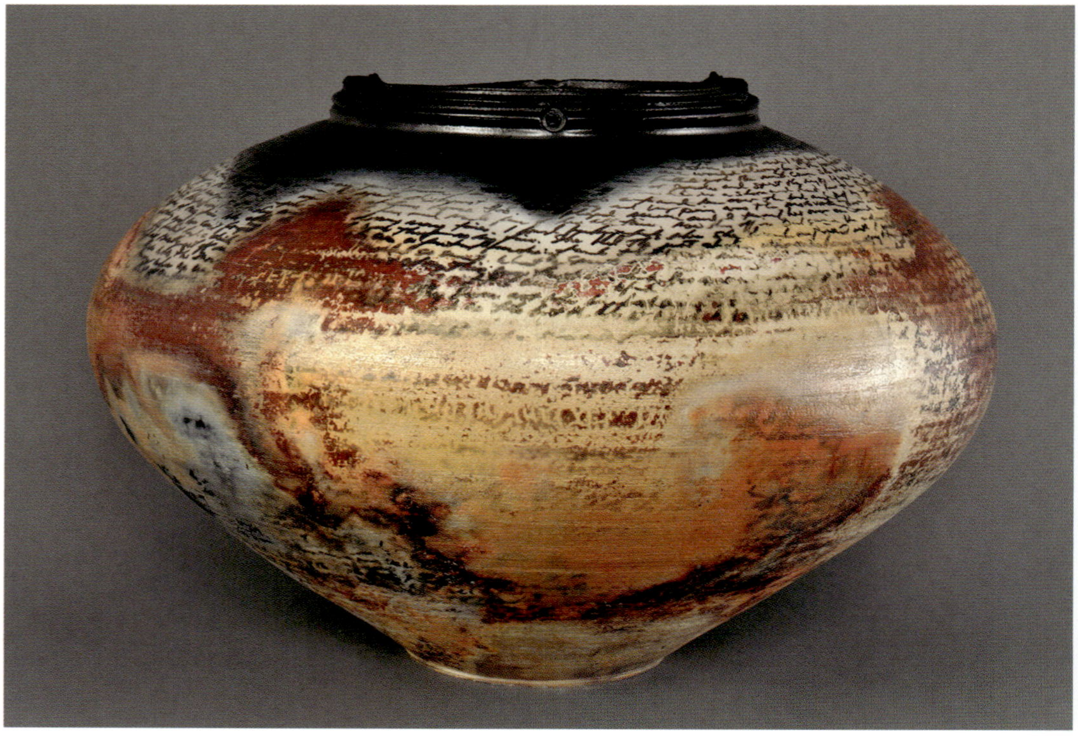

Michael Berkley. *Prophesy: The Return of Nibiru*. Wheel-thrown and altered with white terra sigillata with copper inscription. Copper carbonate, Nick's special dung with holes, salt, sea salt, and seaweed. Cherry, walnut, and madrone wood coupled with Berkley's specialty mix sawdust. 13" h x 19" w. *Photo: Christine Ferrell*

CATHIE JO BUHLERT
(US)

While looking at photos of my work, my new friend Bushra asks, "You make these beautiful things out of mud?" Yes, I create beauty out of mud. I create beauty from the flesh of Mother Earth. I cannot imagine a greater gift than this simple, humble relationship. To fire these ceremonial objects with sticks surrounded by the beauty of the natural world is just about as good as it gets. The simplicity is divine.

I have worked for over thirty years to master the clay, and in the process I have become a servant to the spirit of clay. With every series of pots I create, I find myself falling in love all over again. I find the simplicity and humble nature of the pinch pot very satisfying. My work is guided by spirits and imbued with the spirit of the specific energy I am working with at the time. A Badger medicine pot carries the spirit of Badger in the way a battery carries an electrical charge.

The hours I spend on one ball of clay allows me to experience the process moment by moment: intimately feeling all of the changes the clay goes through on its journey from a moist ball of clay to the moment the burnish is complete.

The making and firing of my ceremonial pottery is in a certain way the center of my spiritual practice. The creation of a ceremonial pipe is the center of that center. The pit firing process allows me to continue the intimate nature of my work with the clay. Offering the raw pots into the fire in this way is an act of faith. Laying the pots gently into the Earth is an act of love. Spending the day or night out in the desert, collecting the sticks for fuel, tending the fire, and finally receiving the gifts of Mother Earth and the spirit of the fire is well worth the blind bargain of the pit firing process.

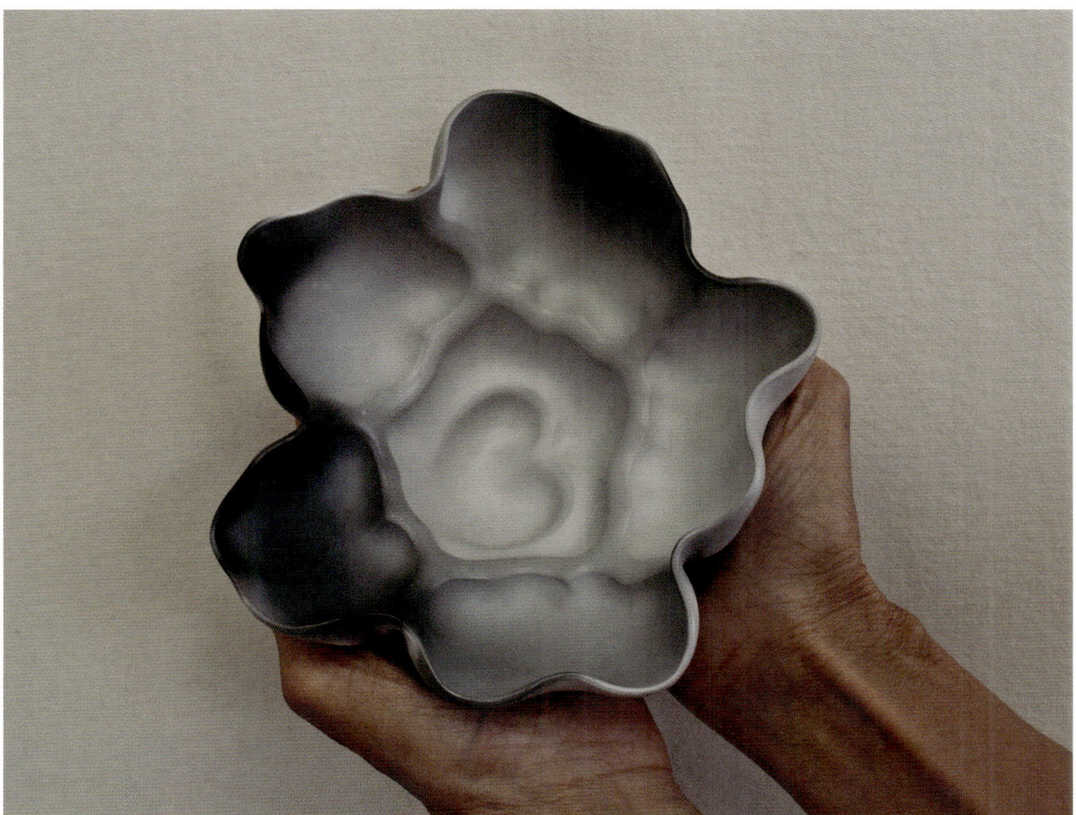

Cathie Jo Buhlert. *Badger Medicine—The Flower of Continuous Bloom.* Handbuilt pinch pot, burnished. 3" x 7" x 6". *Photo: Abigail Oberg*

Cathie Jo Buhlet. *Snake Medicine.* Handbuilt pinch pot, burnished. 2.5" x 5" x 4.5". *Photo: Shannon Sullivan*

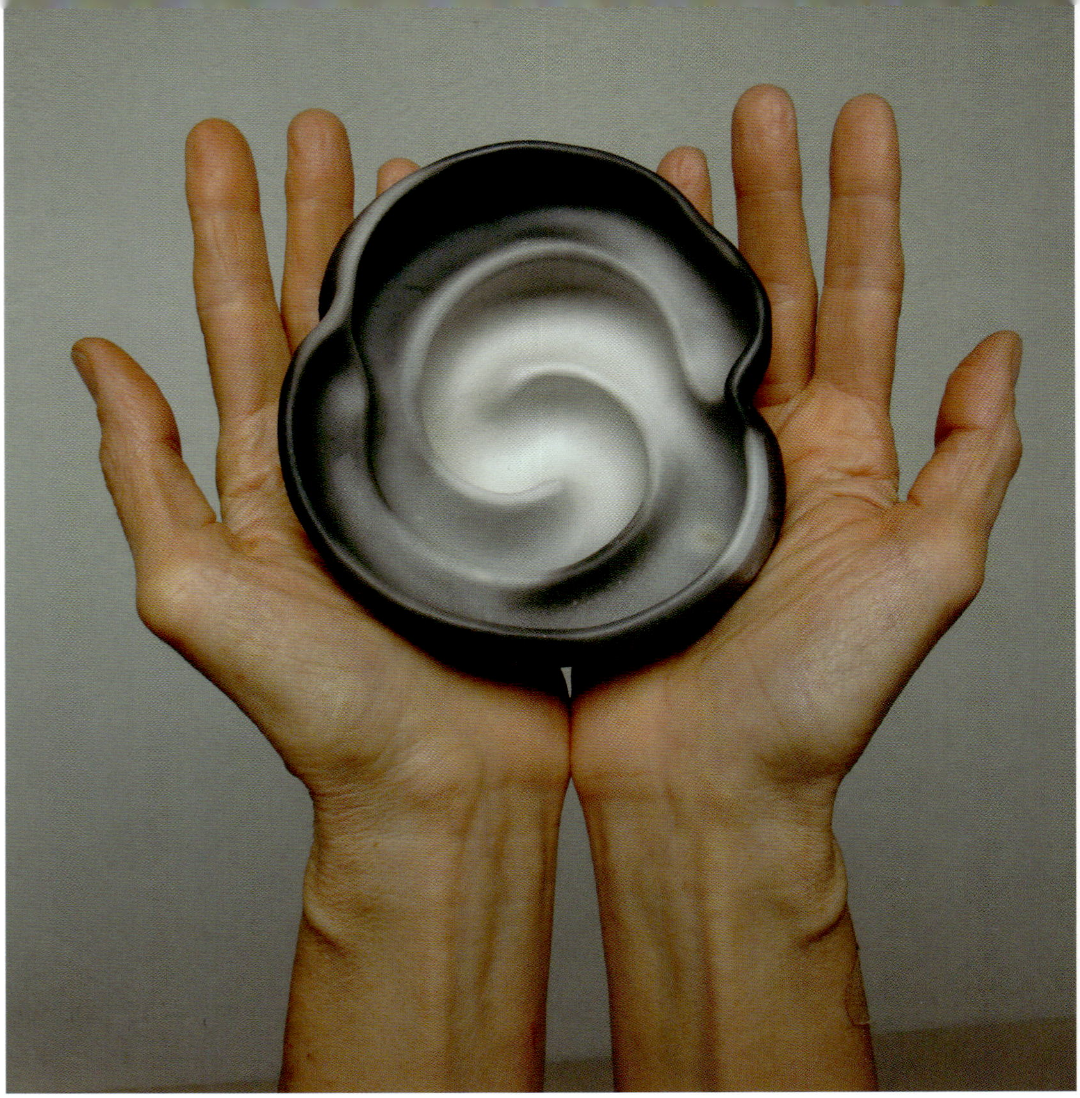

Cathy Jo Buhlert. *Turtle Dances Herself Out of Her Shell.* Handbuilt pinch pot, burnished. 2.5" x 6.5" x 6". *Photo: Abigail Oberg*

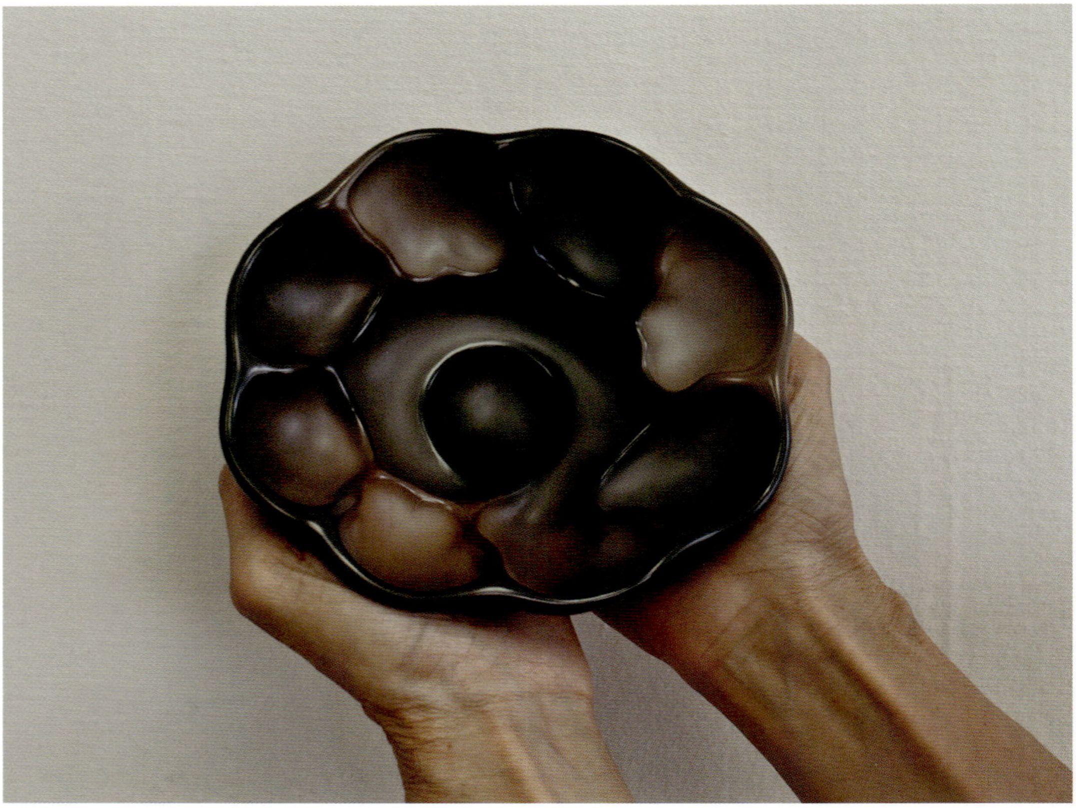

CATHIE CANTARA
(US)

Cathie Cantara is a full time studio potter, specializing in high-fired functional stoneware, as well as raku and pit/saggar-fired art pottery.

Cathie's parents met in art school and she grew up in an environment that nurtured and encouraged creativity. It was while taking a pottery class her senior year of high school that her parents purchased a potter's wheel for her and a seed was planted.

Throughout her adult life, Cathie has always followed artistic pursuits, which included working in a stained glass studio and pursuing her love of oil painting. In 2000, Cathie returned to clay with a renewed passion and began her self-directed exploration of the medium. This culminated in her opening the doors to Homeport Pottery Studio in 2005.

Since then she has participated in group shows with Southern Maine Clay Guild and has been juried into national exhibits and numerous fine craft shows throughout New England.

Of her art practice Cathie says,

I have a fundamental belief that the world around us is enhanced by instilling beauty into the objects that people use and see on a daily basis. While I love making bowls and dishes, I have a feeling of awe when it comes to pit firing. There is a primitive connection to the elements of earth and fire. There is a sense of letting go when I place my work in the pit and light a match to the twigs and seaweed surrounding it. I am continually amazed by the subtle variations of color and shadows on the pots as they emerge from the ashes. This is the work that ties me to the beautiful landscape that surrounds my studio here in Maine—the rugged seacoast and woods behind my house.

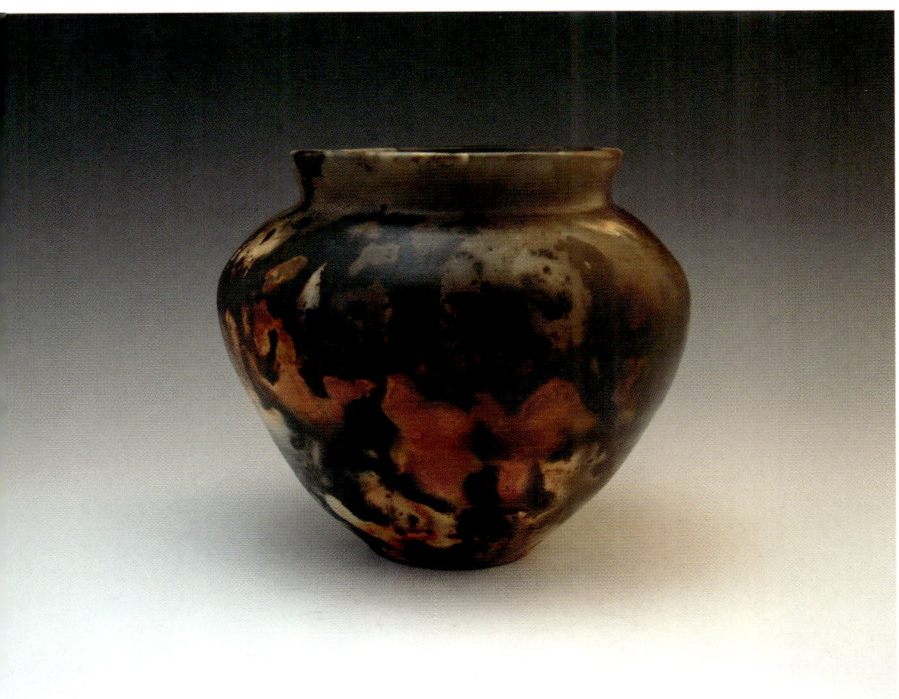

Cathie Cantara. *Cantara 2.* Wheel-thrown, burnished. 6 1/2" x 6 1/2". *Photo: Cathie Cantara*

Cathie Cantara. *Cantara 4.* Wheel-thrown, burnished. 7 1/2" w x 7" h. *Photo: Cathie Cantara*

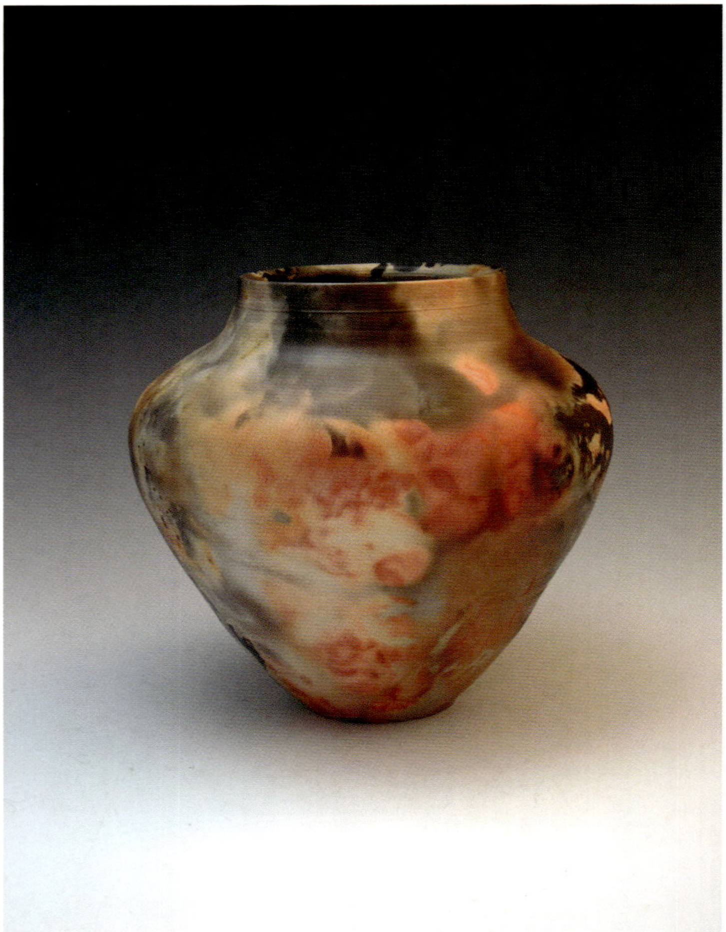

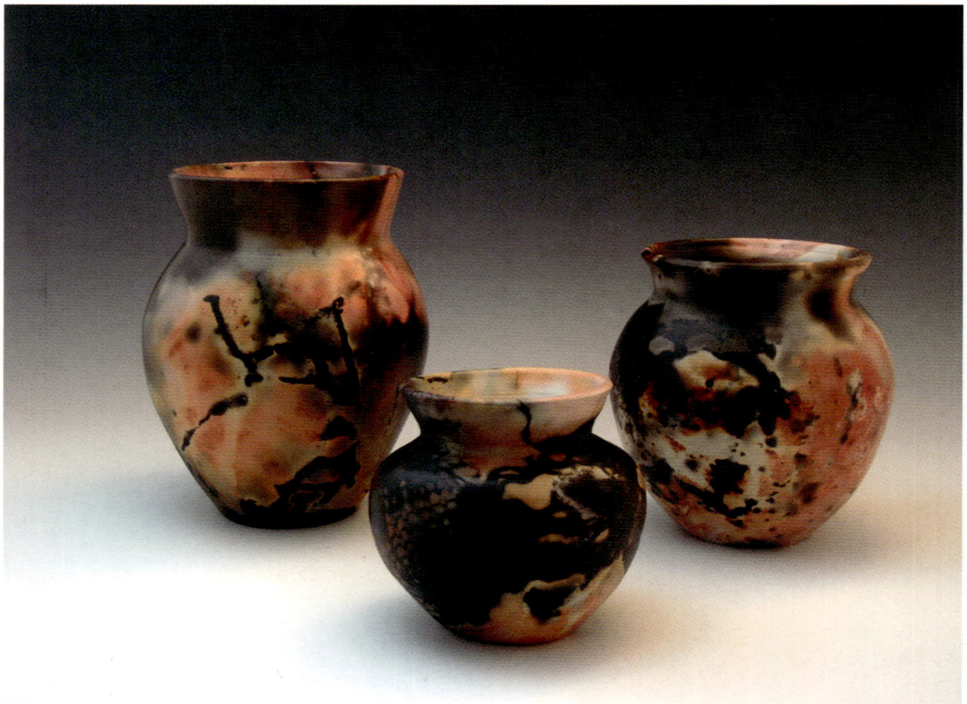

Cathie Cantara. *Cantara 6.* Wheel-thrown, burnished. 4 1/2" x 4 1/2"; 6 1/2" h x 6" w; 8" h x 5 3/4" w. *Photo: Cathie Cantara*

HILARY CHAN
(US)

Though I have worked with figurative sculpture and functional pottery in the past, today my artistic endeavors are concentrated almost entirely on pit/smoke-fired ceramics, with a special emphasis on experimentation and techniques.

I use a simple rock form for all my pieces, each inscribed with a unique serial number, to better identify and correlate experimental results. These clay rocks constitute the preferred form I use for nearly all my pit firing efforts.

In general, the making of these rocks involves two distinct but related processes. The initial formation is done by first hand forming two small pinched bowls, which are joined into a ball when they become firm enough to handle. The ball is then squeezed, slapped, and paddled to take on the shape of a river rock. Finally, as the rock is about to leave the leather-hard stage, it is burnished with a pebble, and then optionally polished after terra sigillata is applied. If no terra sigillata is applied, two or more burnishings will take place to give the rock's surface a high sheen.

This is a very slow, laborious, and meditative process, which I've found soothing and hypnotic, as if I had entered a timeless world shared with my distant ancestors, a world at once simple and yet complex, where the complexity arises from the need to pay close attention to the nuanced response of the drying clay under my fingers, and not from the myriad of accumulated knowledge and human-made rules or conventions: in essence, a primal and direct encounter between one and one's natural environment.

Then, after a concession to modern conveniences by bisque firing in an electric kiln, the rocks—now smooth, uniform, and pristine in white, beige, and light brown, evocative of some kind of mysterious, prehistoric eggs—are ready for the second major phase of transformation: the process of pit firing itself.

The firing is usually conducted in a small open pit, and the process and materials used are meticulously documented, both in the form of written notes and in abundant photographic and video records. There is usually a very specific experimental objective for each firing, for example, to compare the use of copper mesh vs. steel wool as a fuming material, or to verify if extra stoking may result in more vibrant colors. The procedures for carrying out these experiments, as well as the knowledge acquired from them, are then published in my blog. The dissemination and sharing of this technical know-how is an integral part of my pit firing practice.

Finally, within the context of our post-modern world, there is the inevitable question of authorship, creativity, and originality. If the colorful fuming patterns on the surfaces of the rocks are to a very large extent the accidental/incidental contribution of the fire, can I still claim "intellectual property" rights to these resulting imageries? Indeed, how much of them is the result of my work, and how much is that of the natural forces that I have collaborated with? And how will "my" results differ from those of someone else following the same procedures? But these will be the subject matters for some other experiments to be carried out in the future ...

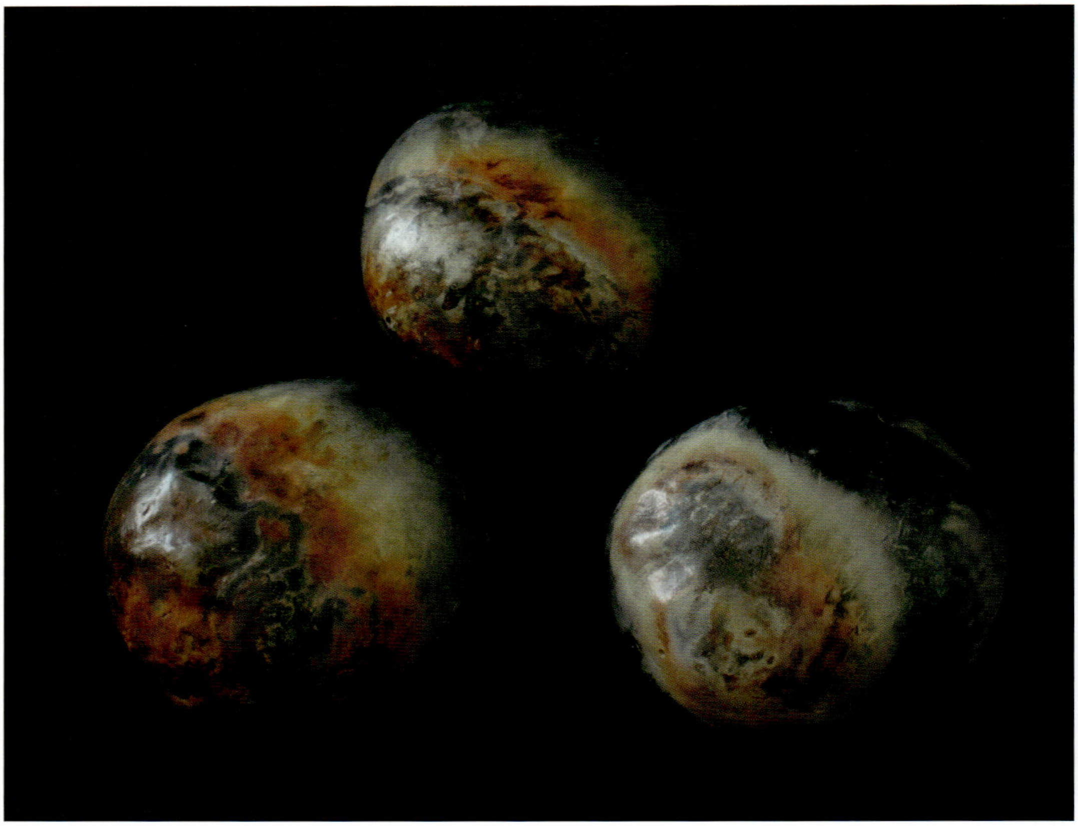

Hilary Chan. *Untitled Group of 3.* Handbuilt, burnished. 3" to 3 ½" diameter x 2" h. *Photo: Hilary Chan*

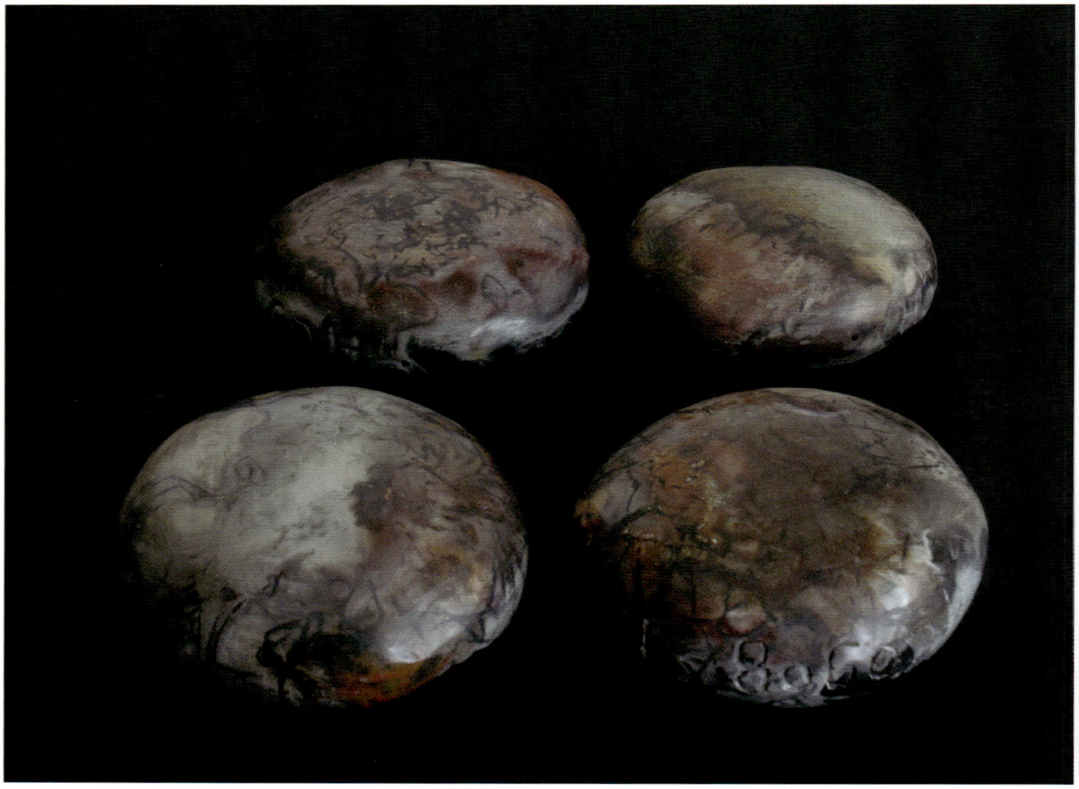

Hilary Chan. *Untitled Group of 4.* Handbuilt, burnished. 2 ½" to 2 ¾" in diameter x 1 ½" to 1 ¾" h. *Photo: Hilary Chan*

MAEVA COLLINS
(CANADA)

I have always felt a deep connection to the ocean. My husband and I spend hours walking the beaches enjoying the ocean and finding driftwood that will enhance my work. I derive great joy in creating my pots and my hope is that others will find equally as much pleasure in viewing these pieces.

There is something to be said for placing your work in a pit and to some extent giving up the ability to control what happens in the next forty-eight hours. We potters make the pots and the flames that dance throughout the pit create patterns and visual magic.

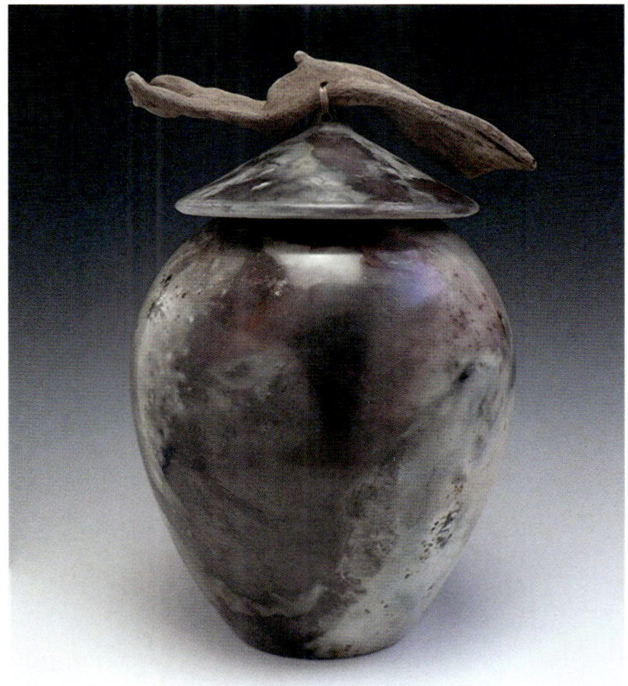

Maeva Collins. *Moody Spirit.* Wheel-thrown and burnished. 13" x 7" [32 x 17 cm]. *Photo: Maeva Collins*

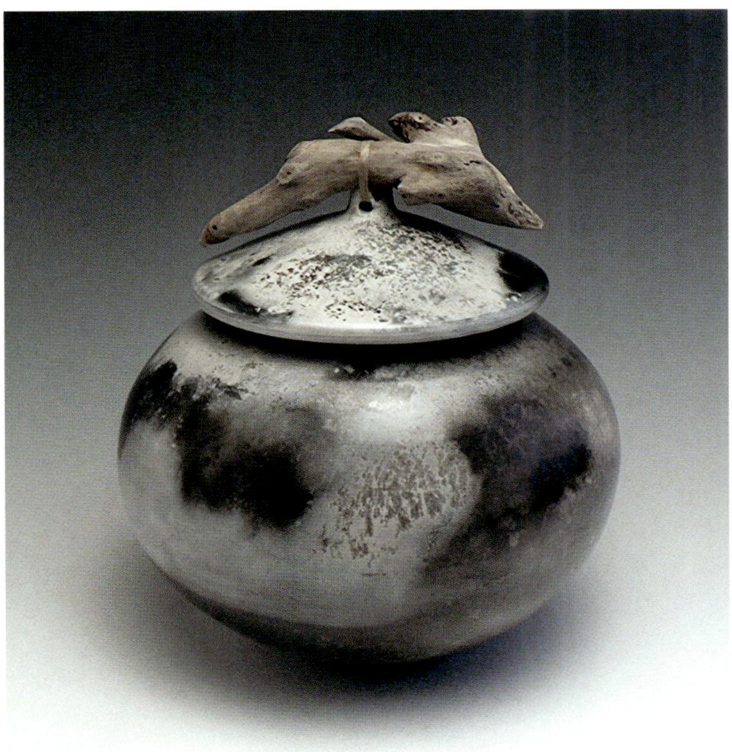

Maeva Collins. *Untitled.* Wheel-thrown and burnished. 7 1/2" [19 cm] h. *Photo: Maeva Collins*

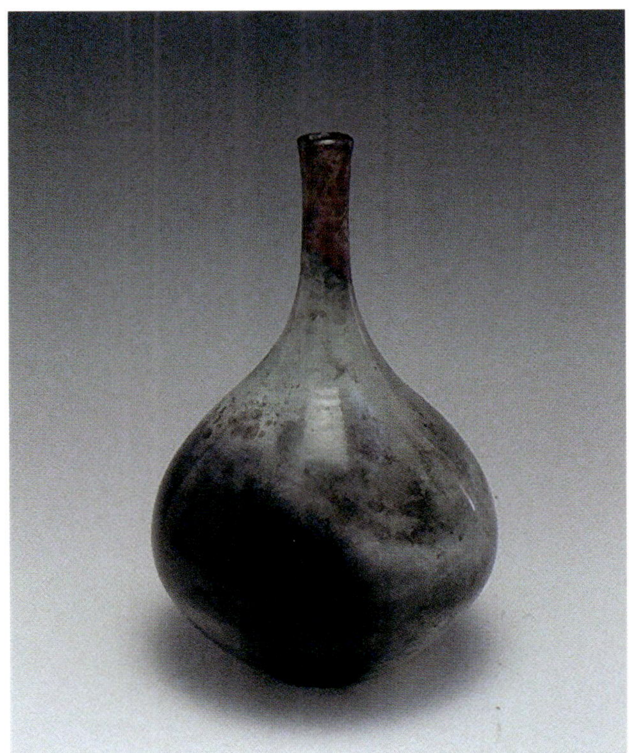

Maeva Collins. *Untitled.* Wheel-thrown and burnished. 7" [18 cm] h. *Photo: Maeva Collins*

ROBERT COMPTON

(US)

After forty years as a working potter, I have found it is important to fire using various methods in order to keep my interest as a potter sharp, and open my work to different markets.

Pit firing is the most primitive firing method I employ, but in many ways it is the most gratifying in the results it produces. While firing to stoneware temperatures in my wood and salt kilns yields durable wares that appeal to a large audience, pit firing gives my work a painterly feeling that I cannot achieve in any other way. The interaction of vaporized metals in the fire on the surface of pit fired vessels results in unexpected patterns and effects that are unique to each and every firing.

After working diligently to control glazes and the temperature of high fire kilns, it is a joy to let loose and give the pit kiln the freedom to mark each pot differently.

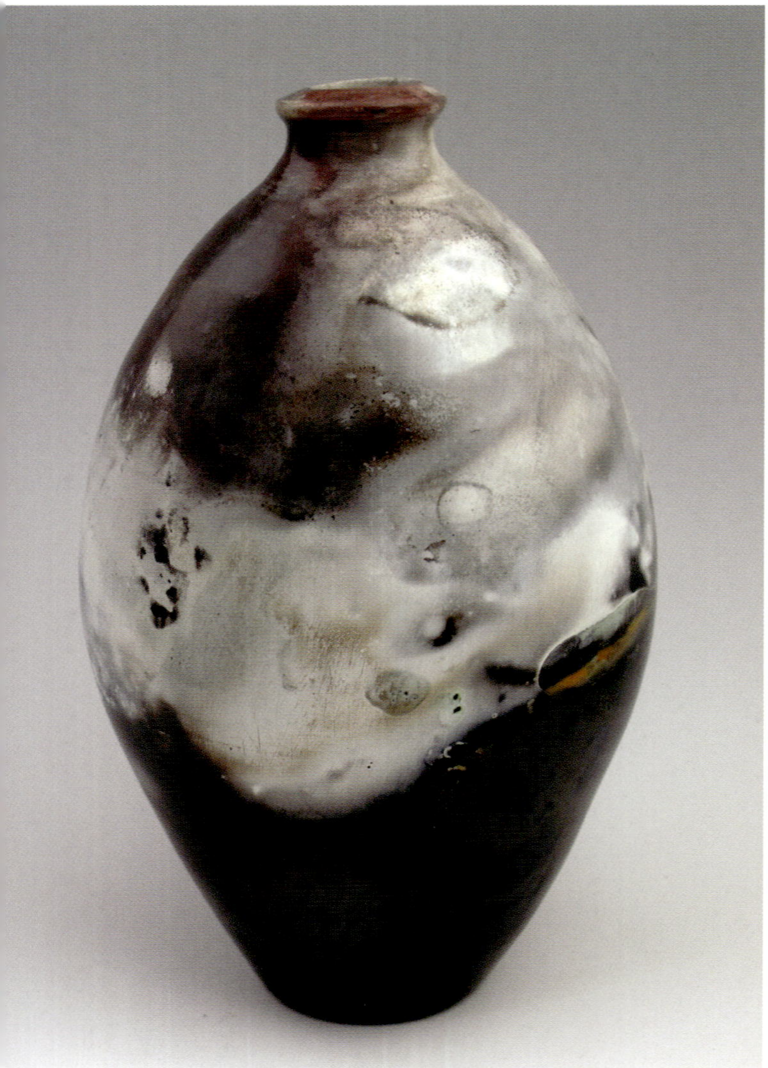

Robert Compton. *Vessel 0240.* Wheel-thrown, pit fired vase, copper fumed and carbon trapped. 9 ¾" h x 5 ½" d. *Photo: Robert Compton*

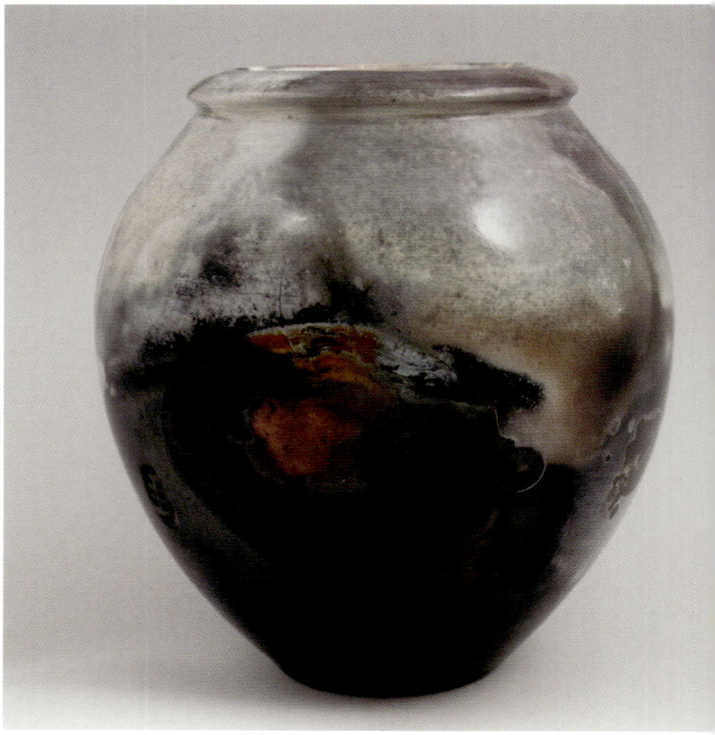

Robert Compton. *Vessel 0034.* Wheel-thrown, pit fired vase, copper fumed and carbon trapped. 6 ½" h x 6" d. *Photo: Robert Compton*

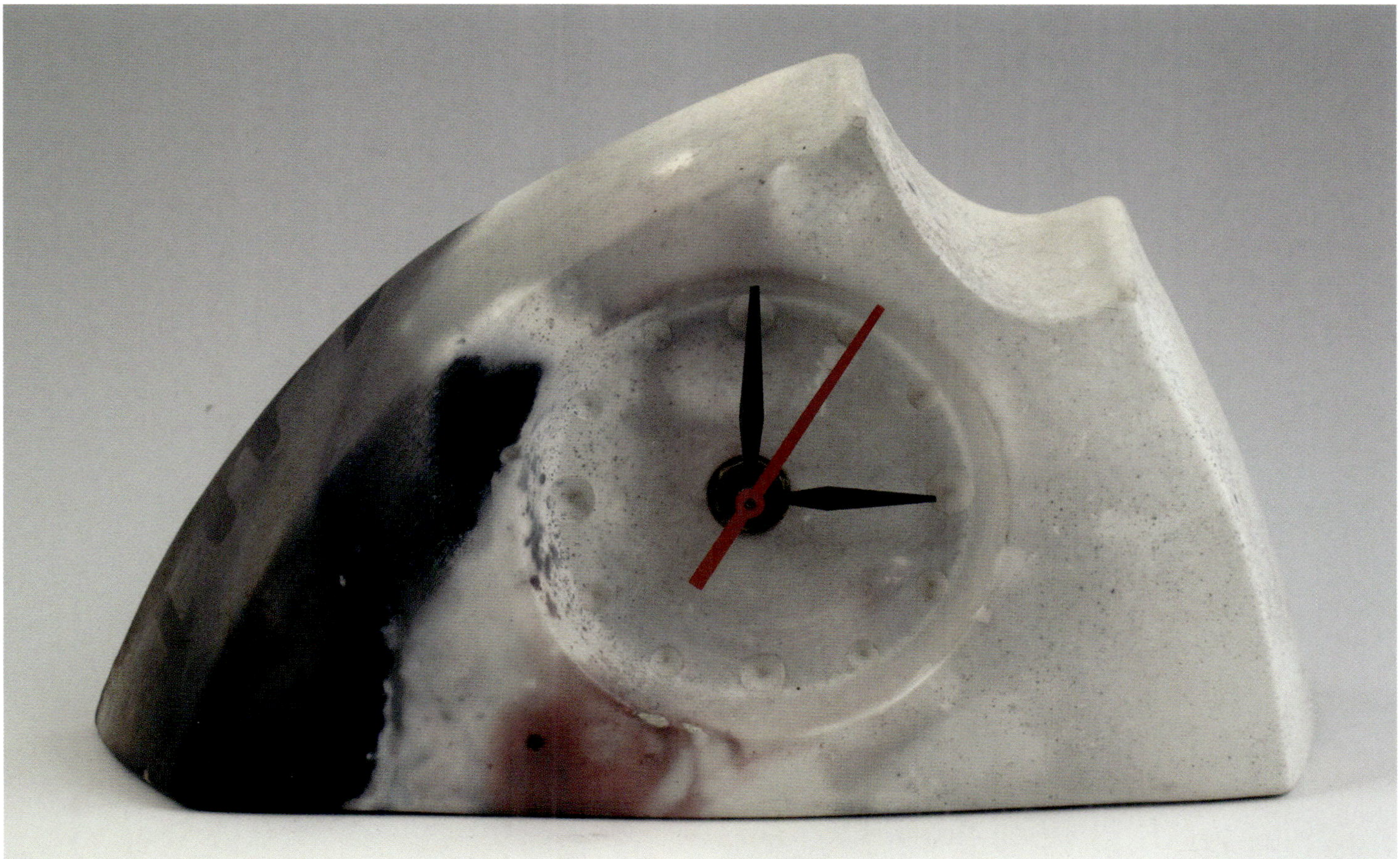

Robert Compton. *Slide Clock.* Slab built, then slip cast. 6" h x 10" d. *Photo: Robert Compton*

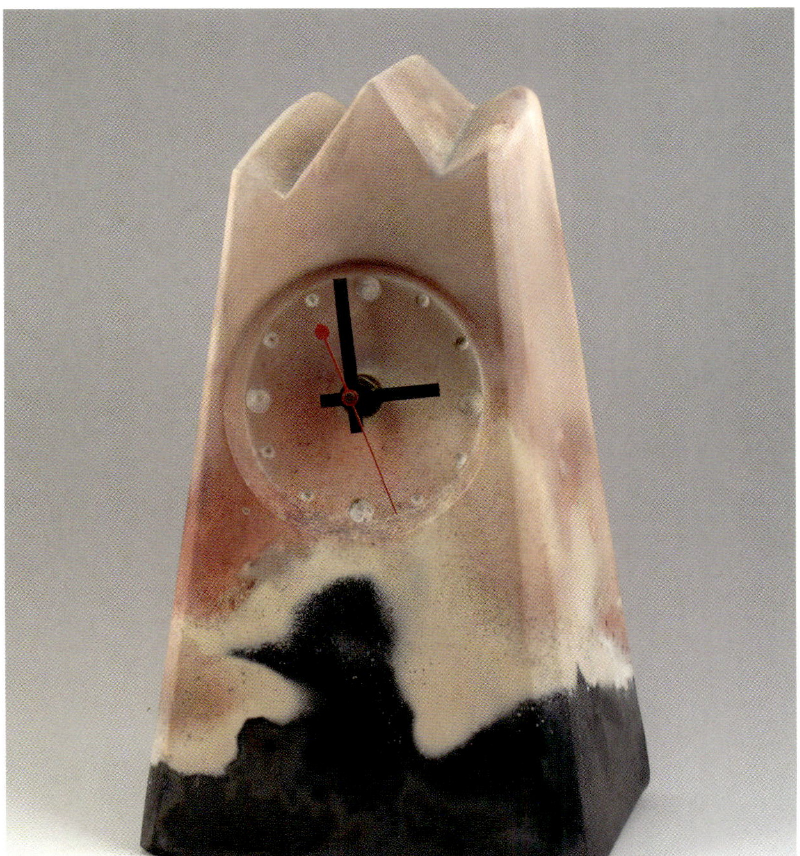

Robert Compton. *Empire Clock.* Slab built, then slip cast. 12" h x 6" d. *Photo: Robert Compton*

ROB DREXEL
(US)

With the broad, beautiful expanse of the Pacific Ocean as a backdrop, my studio partners and I dug our first pits in the soft, white sand of the San Francisco beaches. Packing up our beat-up artists' cars with bisque pottery, wood, sawdust, and a few simple "chemicals" (salt, copper carbonate), we set out just after sunrise from the city workshop for our all day pit firing adventure at the ocean side.

Thus began my twenty year exploration of this fascinating, ancient firing method. Although our ancestors pragmatically prepared pit fires to increase the functionality of dried clay, I sought beauty and diversity of color and markings. This was my path—to find out how to burn a pit for the most brilliant and dynamic earth tones a carefully controlled fire could deposit on my decorative pottery forms. I quickly learned that the processes and method used to achieve functionality (by the ancients) and the procedure needed for optimum colors were very different indeed.

Controlling the balance of oxidation/reduction in the pit, the way the wood, sawdust, and pottery is loaded into the pit, the timing of the added salt and copper carbonate are all specific to achieving the brilliant reds, golds, yellows, blacks, and the diverse blends of all these colors that I most desired on my pottery. I also learned that the preparation of the pottery from greenware to bisque influenced the results I obtained dramatically.

Even after twenty years, each pit is both different and a surprise—opening it with the same excitement as I did my presents on Christmas morning as a child. I'm quite sure the adventure of this process will never end.

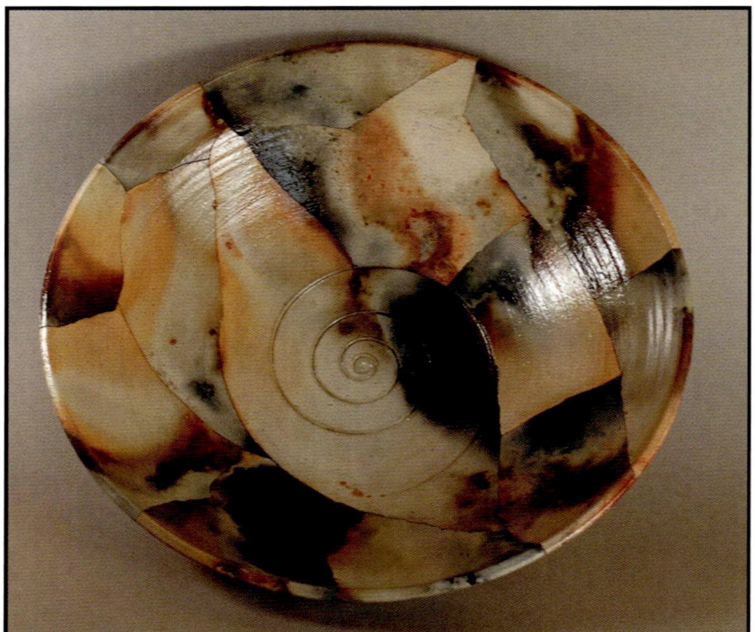

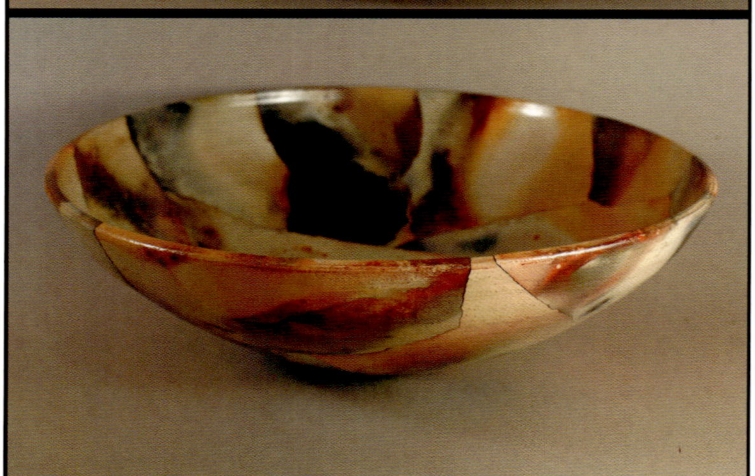

Rob Drexel. *Large Mosaic Pit Fire Bowl.* Wheel-thrown hand-built additions. 19" x 6". *Photo: M. Conway*

Rob Drexel. *Mosaic Lidded Vessel.* Wheel-thrown hand-built additions. 11" x 10". *Photo: M. Conway*

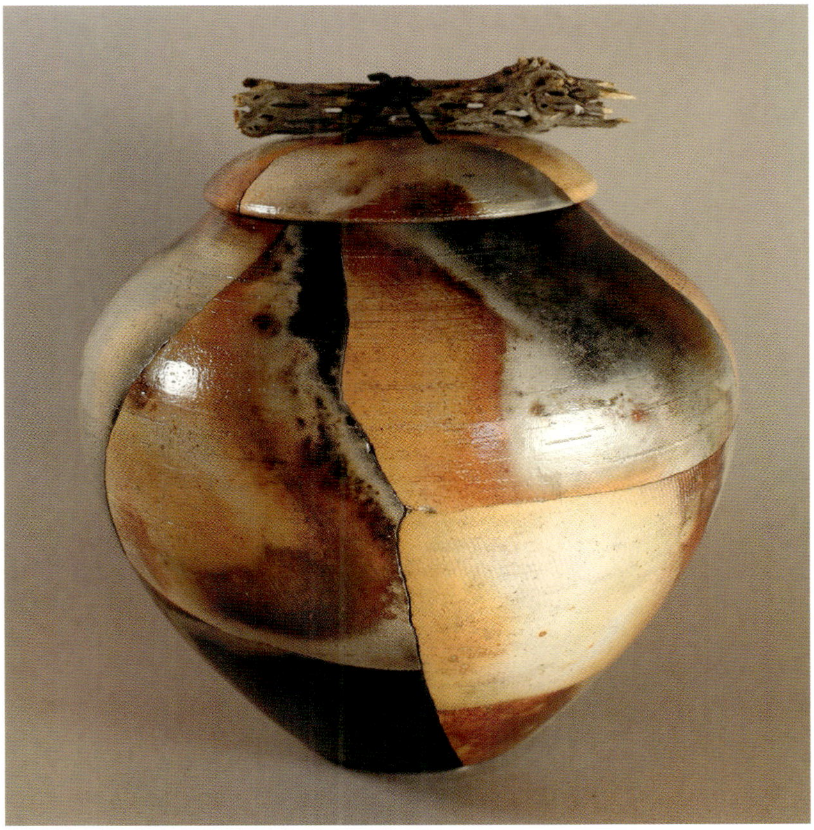

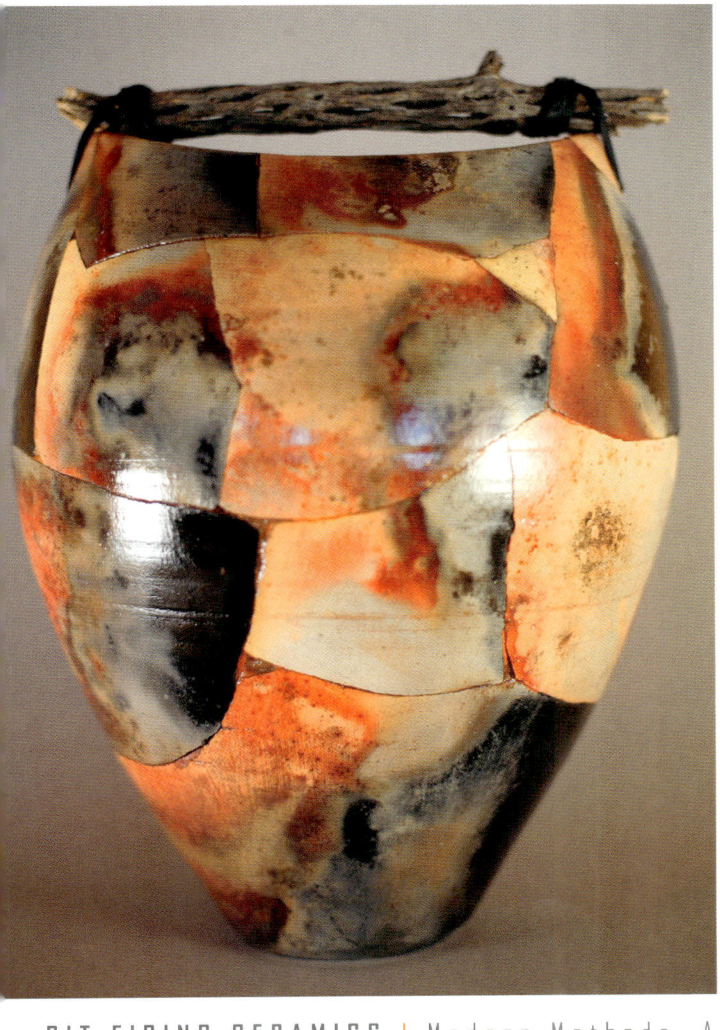

Rob Drexel. *Mosaic Wood-Top Vase.* Wheel-thrown hand-built additions. 13" x 10". *Photo: M. Conway*

SUSAN DUHAN FELIX
(US)

Susan Duhan Felix has been a serious ceramic artist for the last fifty years, specializing in ritual objects. During this period, she has exhibited her work widely in both group and solo shows throughout the USA and has been acquired into several museum collections. Felix has also taught widely in the ceramics and arts disciplines, and has held positions on several art organizations, including the Jewish Arts Community of the Bay, which she founded, and has served as its first Executive Director. Currently, Felix is the Art Ambassador for the City of Berkeley. Felix has won several awards for her contribution to the arts and the community, and was chosen by the International Biographical Center as one of the 2000 Outstanding Artists and Designers of the 20th Century and received their 21st Century Award for this achievement. Most recently, she was chosen to be in the *World Who's Who of Women* for 2009.

Of her art practice, Susan Duhan Felix says:

> My work is pit fired ceramic. Pit firing is one of the most primitive of firing methods, requiring surrender of control over results. I surrender my artwork and control to fire and trust in what will emerge. This process is dependent on trust and faith. Just as one day can be more beautiful than the next, I am given differing blessings from the pit. A few are more amazing than others. As we can trust in the light returning on a daily basis, I have come to trust the powerful pieces emerging from the flames. I am drawn to the mystery and magic of the ancient and try to evoke similar responses in my art.

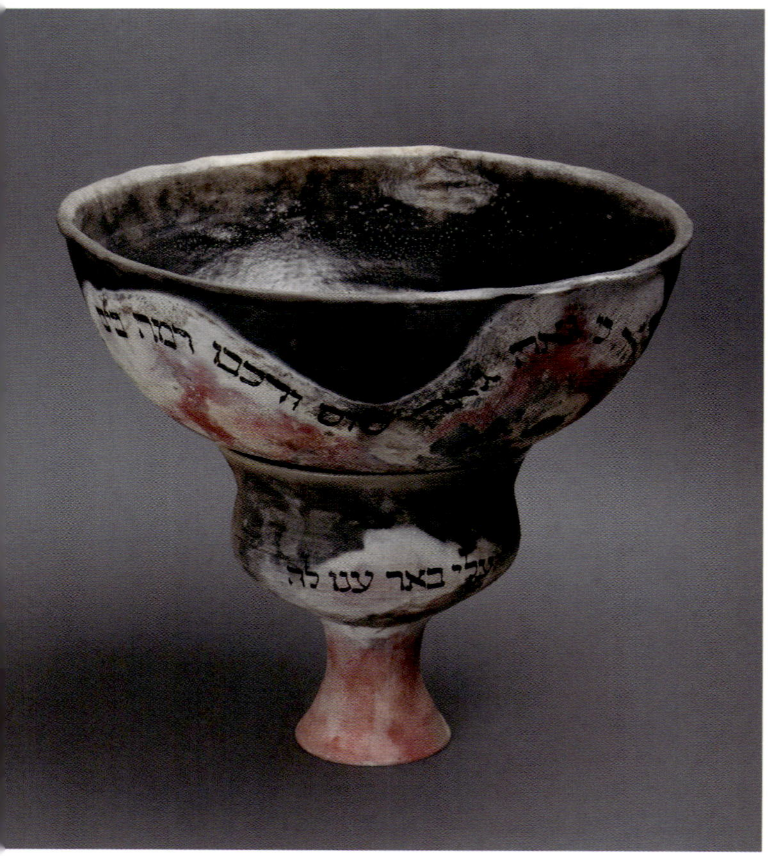

Susan Duhan Felix. *Two Part Miriam's Cup.* Wheel-thrown, pit fired. 11 ½" x 12" x 12". *Calligraphy by Brenda Goldstein. Photo: Oriol Cara*

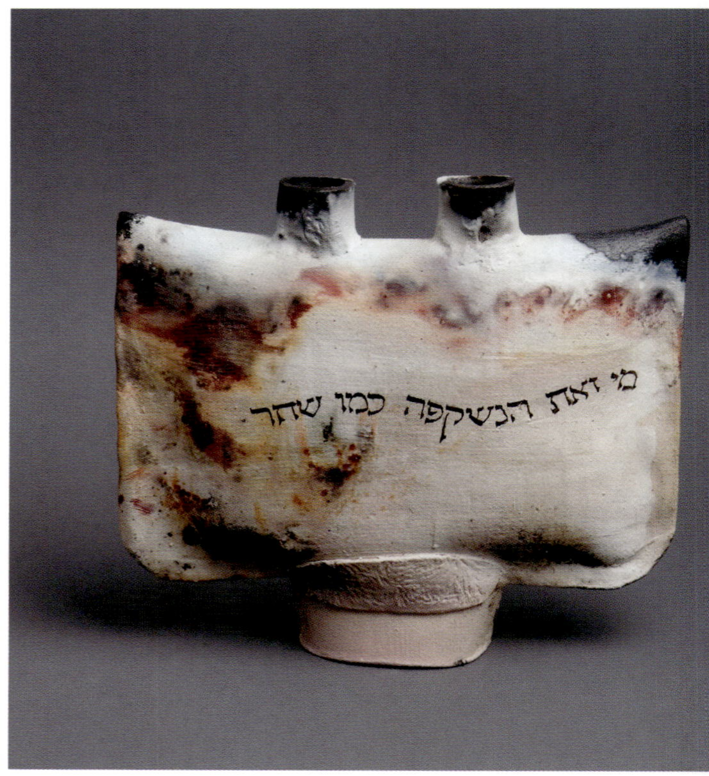

Susan Duhan Felix *Who Is She Who Shines.* Handbuilt, pit fired. 11" x 10" x 3". *Calligraphy by Brenda Goldstein. Photo: Oriol Cara*

KEVEN FRANCIS
(AUSTRALIA)

Keven Francis has been involved in the arts industry for three decades in a wide variety of roles from project management to film direction to visual art. He began working with ceramics in 1981 in Perth, Australia, and continued when he moved to Broome in 1987, where he began an intense investigation into pit firing. Francis completed a Graduate Diploma of Art (visual) in the ceramics workshop at the Australian National University in 2009 and is now concentrating on developing his ceramics career.

Keven Francis says of his work:

> I moved from producing my own glazes and production ware to the excitement and spontaneity of what I call dynamic pit firing. This process excites me as an artist and allows for the fire's spontaneity to combine with the sense of risk that heightens the senses.
>
> The transformative quality of fire has always intrigued me, particularly the intense marks of carbon created on ceramics and wood when immersed in the flames. The theme of transition sits behind the work. Initially this was investigated as a threshold or divide, expressed as the strong edge produced by the fire's carbonization of ceramic and timber. In recent works, this concept is taken beyond the dividing edge to expand the transition zone itself, a separate state between before and after.
>
> My pit firing today is focused on embedding the pit fire's mark on ceramics and paper. My recent work is a collection of several hundred individual pots that are handbuilt, burnished, and then pit fired. Each piece is designed to be individual, with the commonality of fitting my hand and the process of shaping and burnishing
>
> Through the pit firing, carbon is infused into the bodies, then selectively burnt out to reveal a sharp separation between dark carbonization and clear ceramic. They are installed as a collective mound on elevated clear glass with their openings engaging the viewer. The glass creates a platform that evokes ambiguity between their weight and suspension. The pots appear in a state between floating and falling.
>
> The scorched paper works that are created by placing hot pots on the paper are delicate pieces relaying a quiet residue of the fire's intensity. The fine spontaneous charcoal marks sit onto the paper and then appear to float when the work is displayed vertically.

Keven Francis. *Between*, ceramic installation. Pit fired ceramic, Starfire Glass, burnt wood, and paper. 2' 5 ½" h x 2' 7 ½" w x 3' 7 ¼" d. *Photo: Keven Francis*

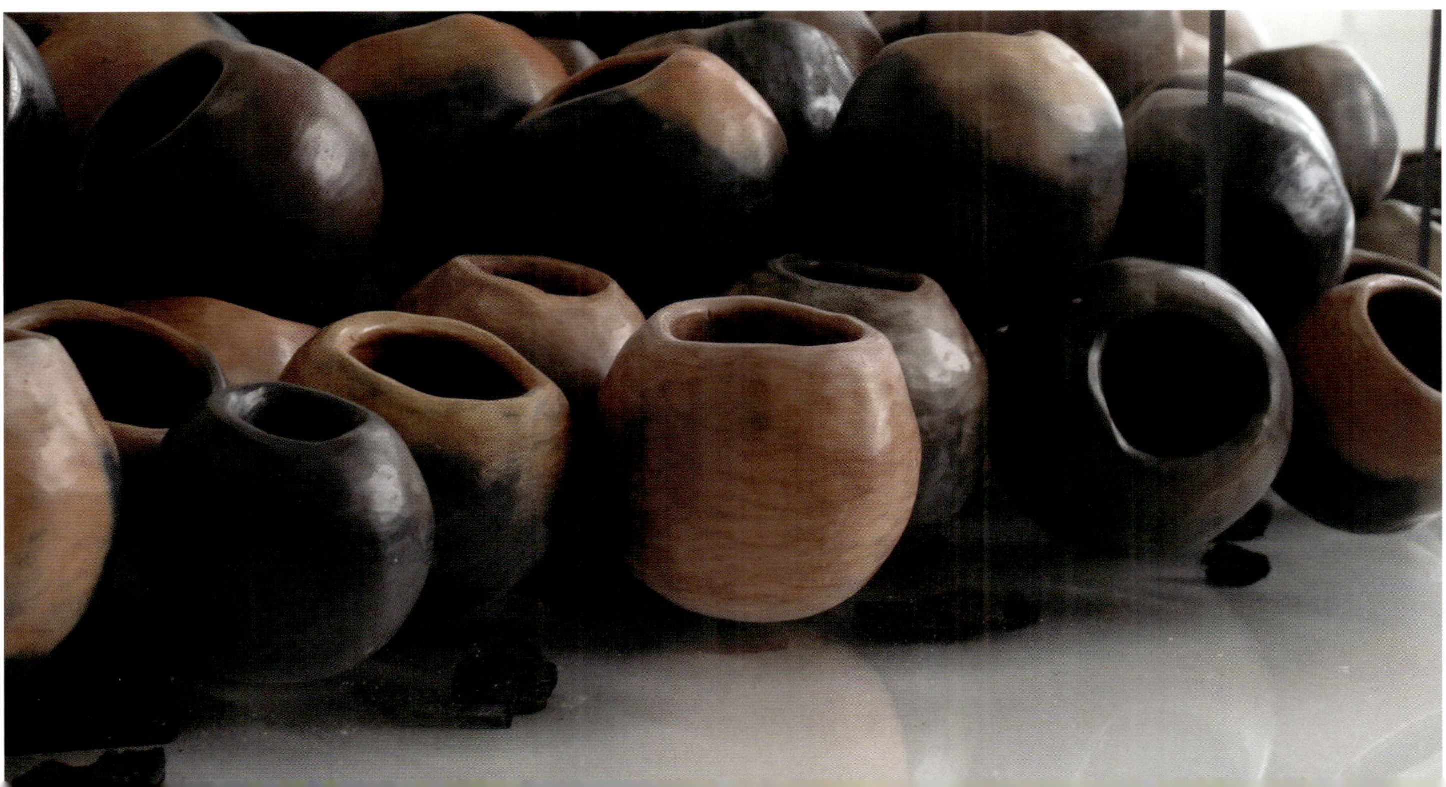

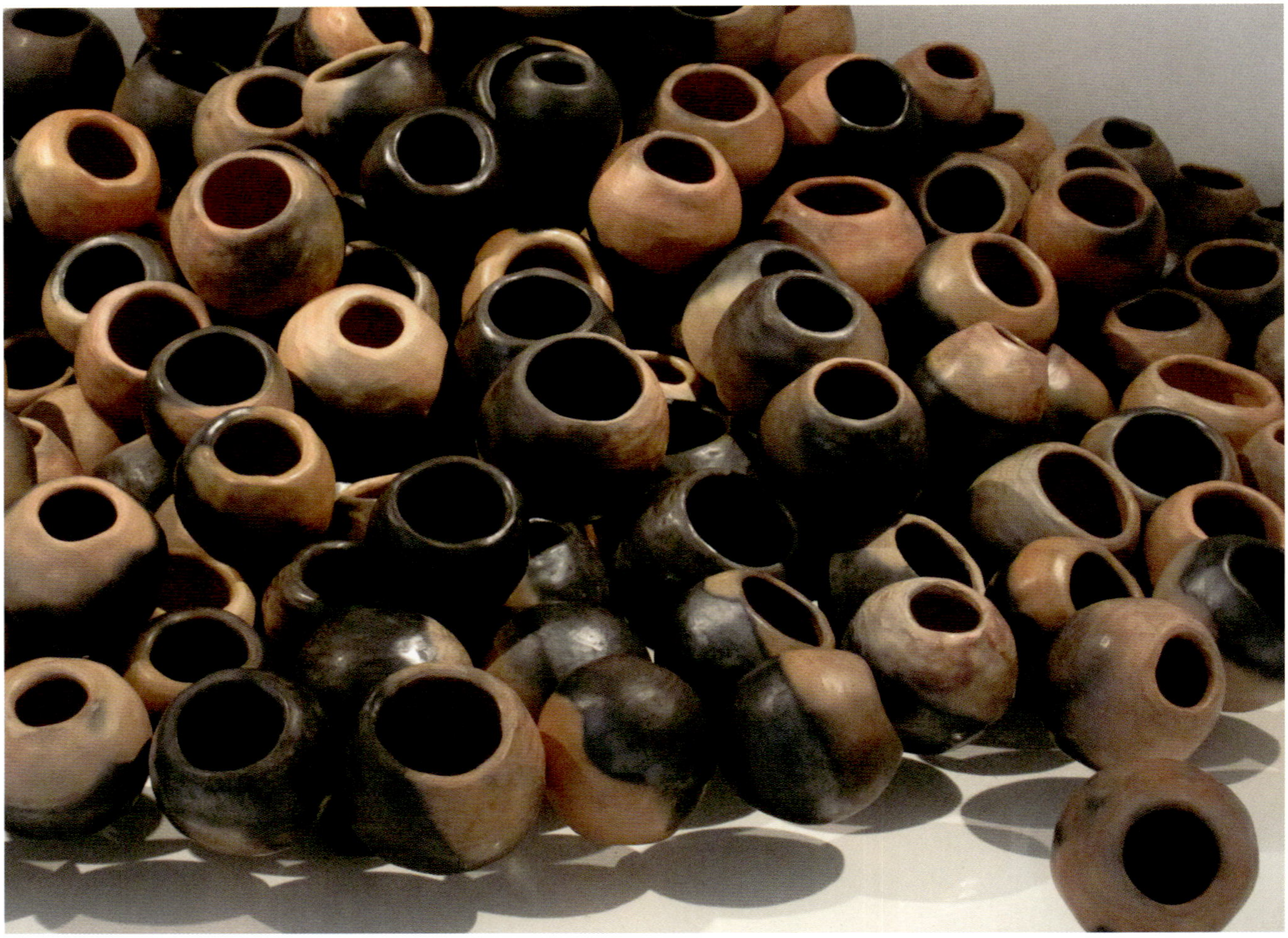

Keven Francis. *Between*, ceramic installation. Pit fired ceramic, Starfire Glass, burnt wood, and paper. 2' 5 ½" h x 2' 7 ½" w x 3' 7 ¼" d. *Photo: Keven Francis*

CHRIS GOLD
(US)

Chris Gold throws white stoneware into shapes on a wheel, burnishes them, and then fires them to 1650 degrees Fahrenheit in an electric kiln to strengthen the work for the next creative phase—pit firing. Gold uses saw dust, wood scraps, salt, and copper to achieve the various mottled tones seen on her pots. When cooled and removed from the ash, Gold polishes them with a coat of paste wax. When throwing, she chooses shapes with a lot of surface area to maximize the impact the colors have on the pot. Gold views the randomness of pit firing as being a reflection on the unpredictability of life and the unique nature of us all.

Of her art practice, Gold says:

> My goal is to have people experience each vessel as a distinctive creation by means of their tactile and visual senses. Burnishing the surface of a pot gives it a satin smoothness that begs the observer to touch its surface. Sometimes the ash will create an area of contrasting coarse texture, making the pot more interesting to hold and turn over in their hands. While I'm buffing them, I experience a sense of giddy excitement as the variable patchwork of colors the flames has imparted on the pots emerges on the surface. The vivid or muted color patterns engage the visual senses so that the viewer never tires of it; they can experience the pot many times and find a new pattern or shade of color they didn't notice before. Some of my most recent work adds subtle details such as slip applications or small carvings. This enables the observer to enjoy the pot from a distance and also offers a sense of surprise when discovering these embellishments on closer inspection.
>
> Working with clay is all about transformations. The Bible tells us we have been created from the dust of the earth, that God molds and shapes us like clay into the people we are. As a potter I don't perform miracles, but I can take a piece of amorphous, pliable material that has little utility and process it into a vessel that is solid, serviceable, and aesthetically pleasing. This transformation of mud to pot reminds me of our origins as humans and never ceases to fascinate me.
>
> Pottery engages the spiritual, emotional, and physical dimensions of my existence and gives me a sense of immense contentment. Any talent I possess in working with clay is a gift from God and I am blessed to be able to create with it. I hope to give others a sense of enjoyment when they experience my work as well.

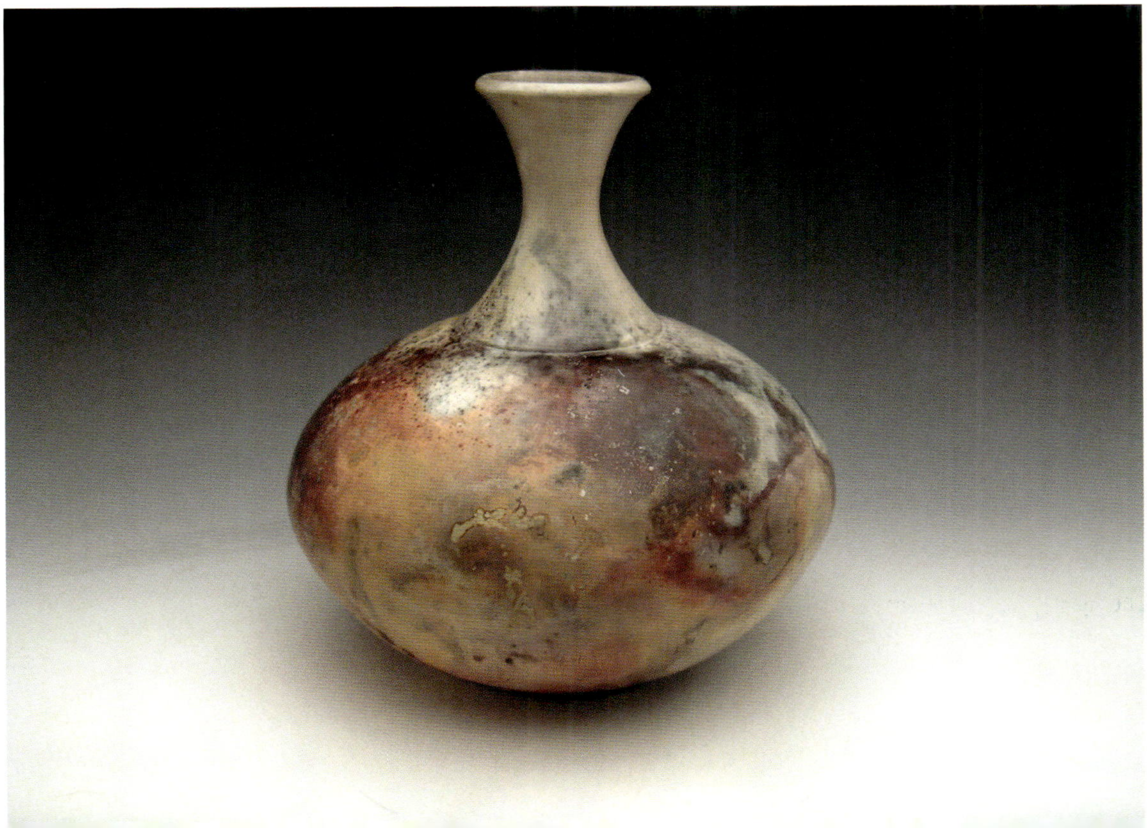

Chris Gold. *Wide Bottom Bottle.* Burnished. 10" h x 9" w. *Photo: Chris Gold*

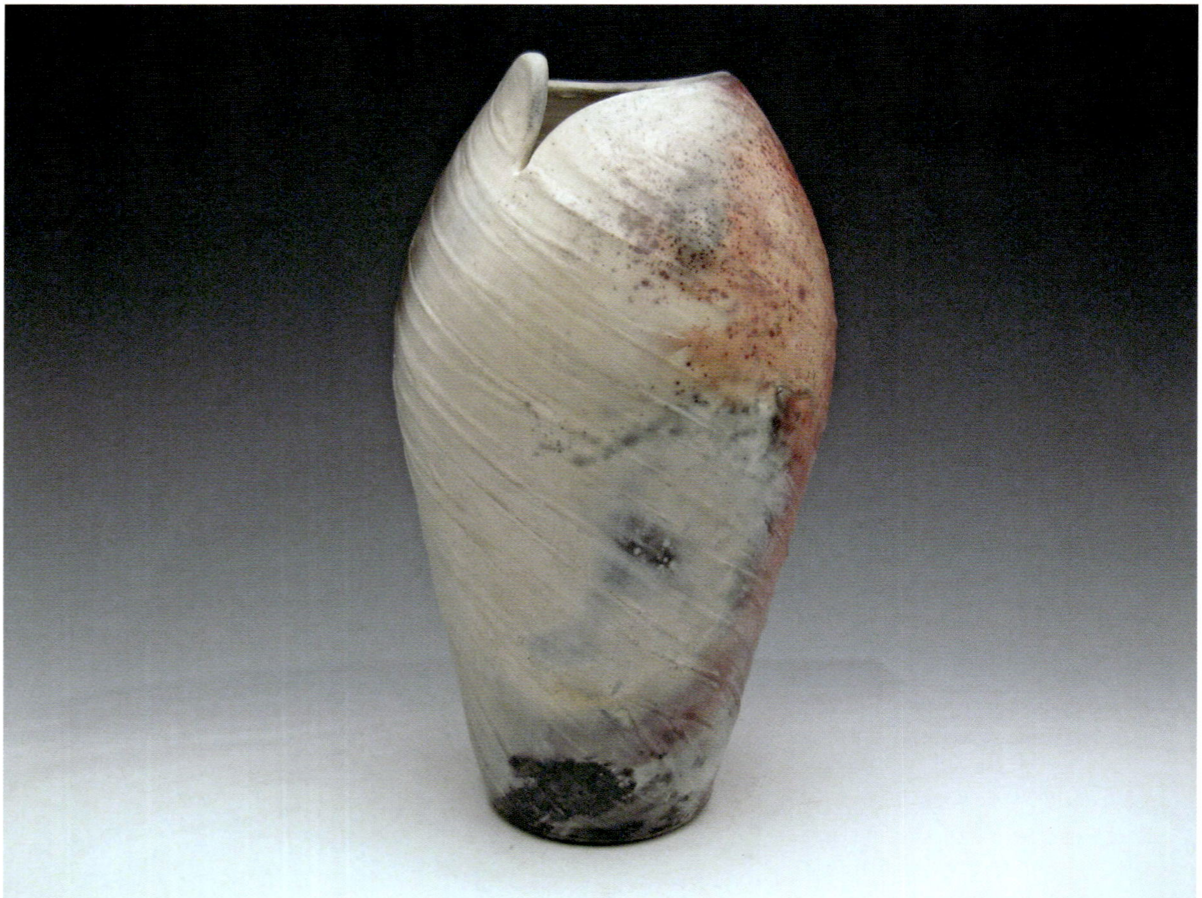

Chris Gold. *Textured Vase, Altered Rim*. Slip coated to add texture, burnished. 11" h x 6" w. *Photo: Chris Gold*

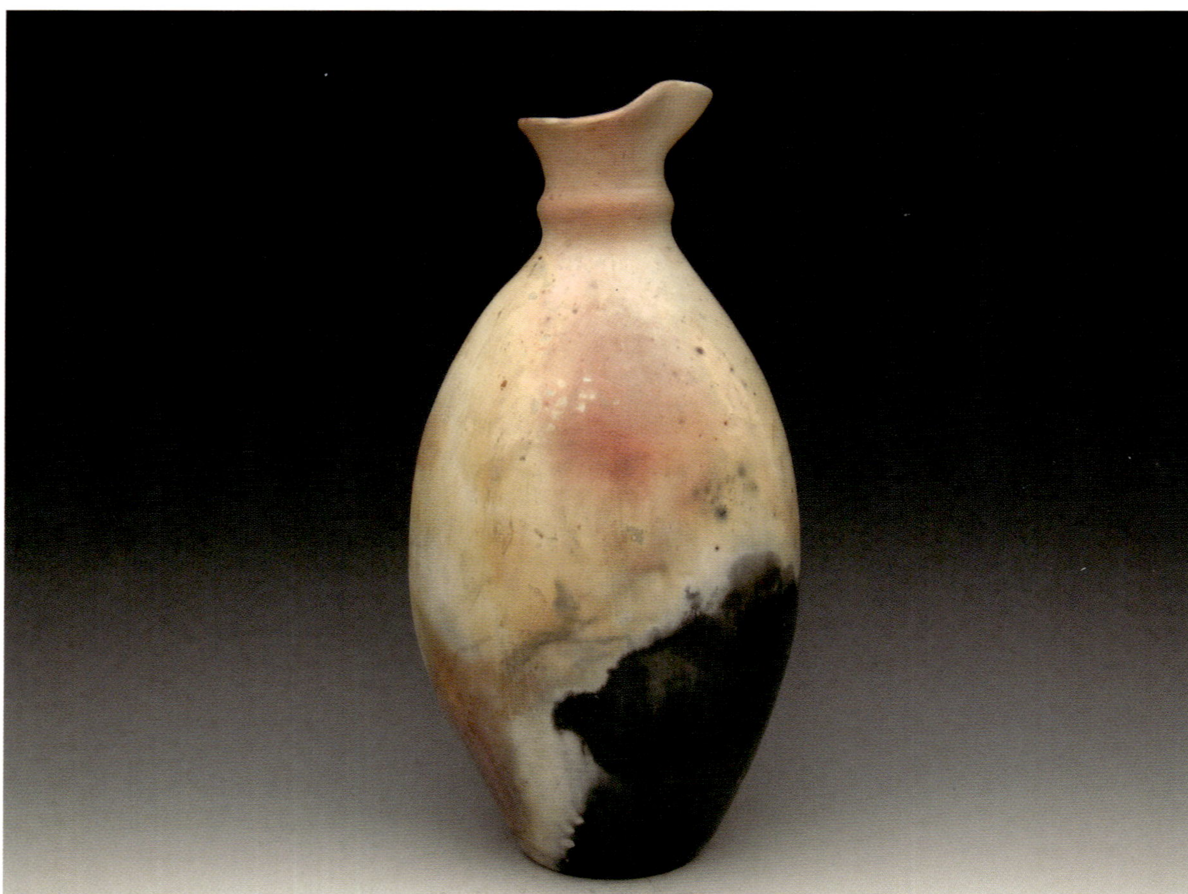

Chris Gold. *Asymmetric Bottle.* Burnished. 14" h x 8" w. *Photo: Chris Gold*

ED GRAY
(US)

Although my hands gather and shape elements of the earth, it is the primal force of fire that completes my work. Earth, Air, Fire, and Water: the four sacred elements that are the breath of life.

These are the gifts with which I give honor to the teachings of my ancestors. It was in 1964 that Chief Little Elk, Mount Pleasant, Michigan, USA, gave me my native name, *Jikiwe* (my friend), so it is with that name that I sign my work in remembrance of who I am and where I came from.

I create my art with a celebration, embracing cultural differences past and present. I weave all into a moving vortex, never to stop creating impressions and staying focused on my visions.

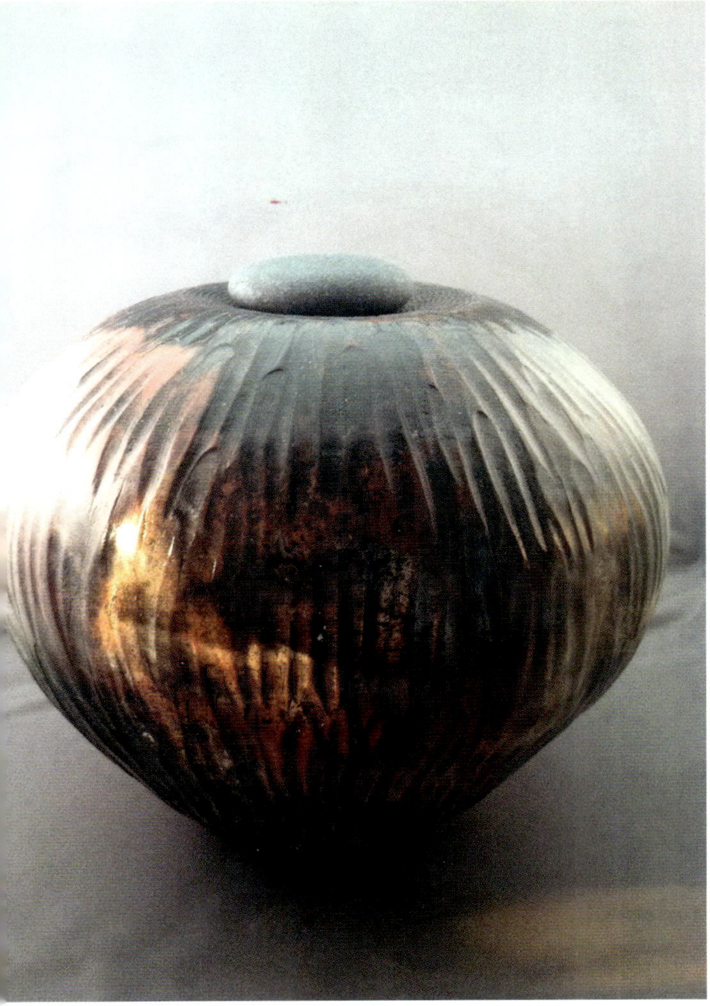

Ed Gray. *Untitled.* Wheel-thrown, carved, pit fired. 10" h x 11" w. *Photo: Ed Gray*

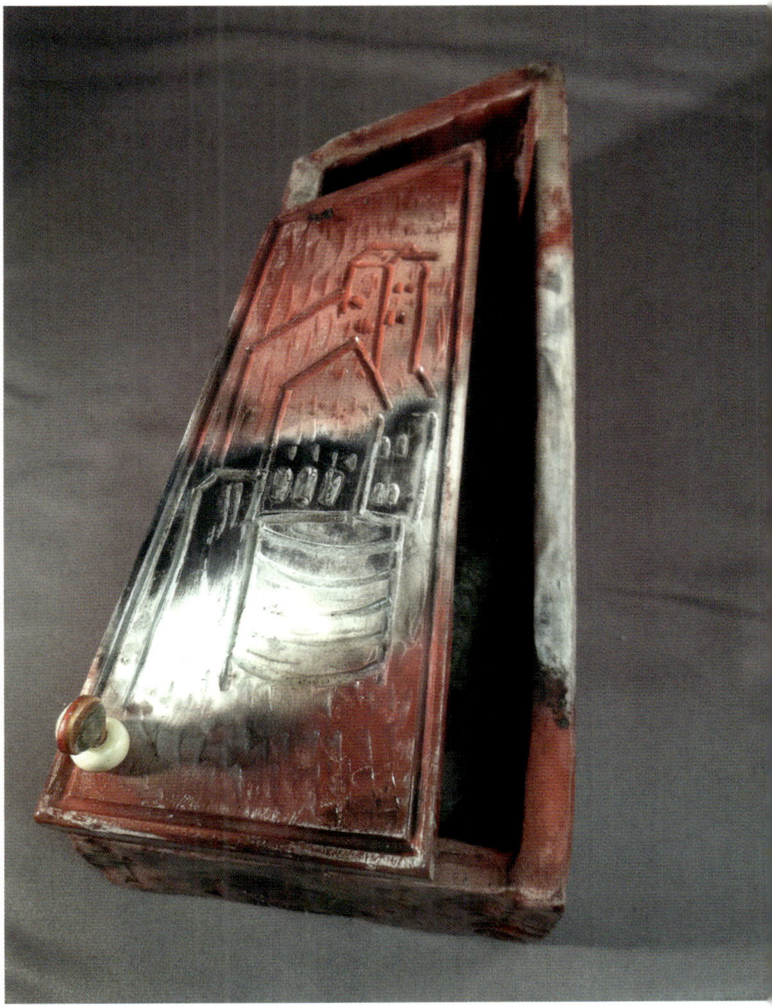

Ed Gray. *Reliquary.* Handbuilt and carved with Osceola Mine Shaft House, pit fired. 8" x 3 1/2" x 3" deep. A piece of copper from the old mine is inside the box. *Photo: Ed Gray*

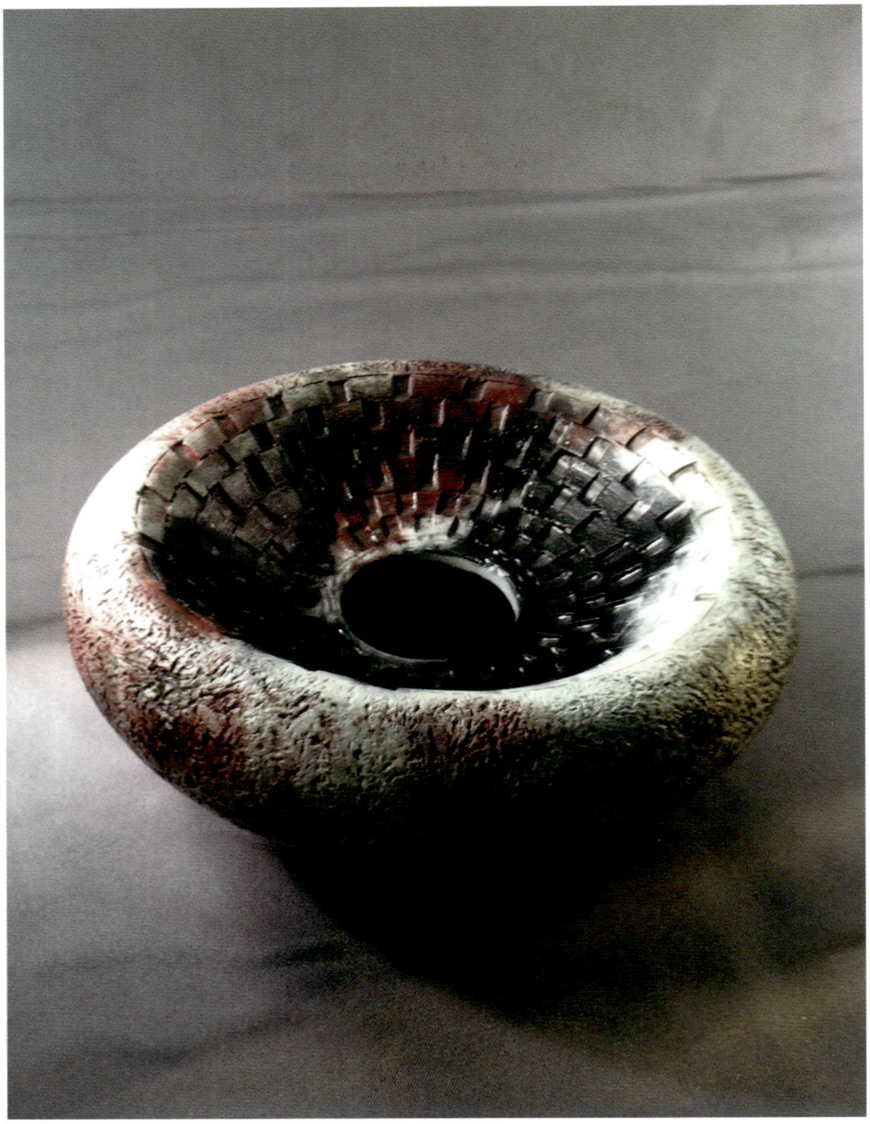

Ed Gray. *Untitled.* Pit fired, wheel-thrown, and carved. 8" h, center hole 2" w. *Photo: Ed Gray*

DAVID HARDER
(US)

Since early childhood, I've always experienced knowledge and understanding through touch. My parents and two older sisters established artistic talents in oils and watercolor. My younger brother developed into an airbrush genius. Until 1999, my artistic talents focused on leather craft and woodworking. After a wheel-thrown pottery demonstration, I was convinced I wanted to try this art medium. Initially, I mimicked other potters whose work I admired, often copying their style almost perfectly. Within two years, expression of my own style and flair evolved.

I admire the classic and ancient forms and place importance on the shape of a piece, often seeking to imitate the graceful lines of the female form. Pushing clay to its limits, I originally turned out extremely thin and fragile pottery, including wide amphora vases that sat on tiny foot rings. Eventually, I settled on stability between the height and width, even weight distribution from the foot ring to the opening, and balance between the empty and filled space of my art. After six years of refining my clay art forms, I was honored when Lark Books, *500 Pitchers*, selected my art from among the 4,500 worldwide entries in 2006.

My favorite clay body is Sonora White by Aardvark Clay. It produces pieces that can be stretched thin or manipulated, yet remain strong enough to withstand the thermal shock of Raku and pit firings. Because the clay remains white and smooth, I can achieve a stark contrast between the glazed area and the smoky black surface. I am beginning to introduce sculpted work into my art, a new and challenging dimension. My pieces range from small Japanese sake sets to large Amish bread bowls, to sizeable jars/vases, 60" tall. Maintaining the form throughout the throwing process of a tall piece can be demanding, especially when joining the large rings together.

When decorating my art, I enjoy colors and surfaces that are found in nature. Raku was my first experience, but when I tried pit firing, the random and uncontrolled results were impressive, almost alive. Surface preparation requires the application of Terra Sigillata, watery silt containing extremely fine clay particles, and burnishing each piece, demanding patience. I've instructed several workshops from basic throwing techniques to pit firing and across a wide range of age groups. Personally, I find these activities astounding and rewarding.

My current work gathers inspiration from the desert where I live, including the delicate hues found during spring and fall. Branching out, I will incorporate more sculpture into my work and include architectural pieces. As always, I seek to make each piece unique. I am delighted to know that my work is placed and selected by my customers because my art "spoke" to them.

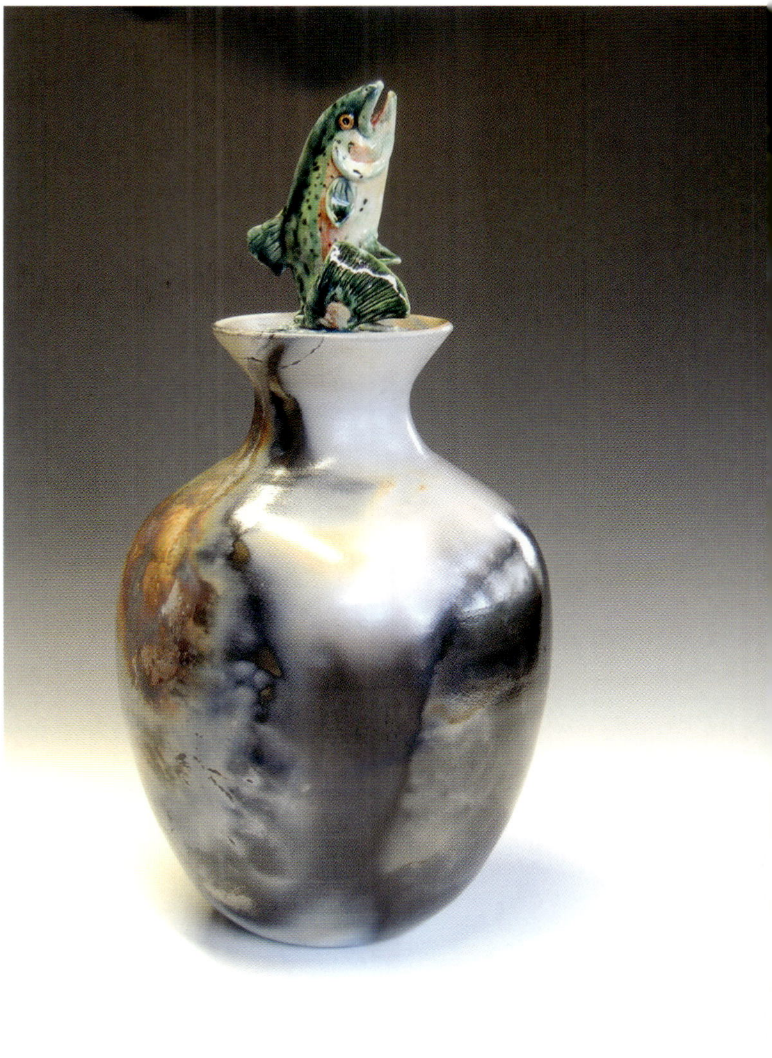

David Harder. *Jar-Lid*. Jar wheel-thrown using Aardvark-Sonora White clay, burnished, Terra Sigillata, buffed with a plastic grocery bag three times. Lid—hand sculpted from paper clay, hand-painted using low-fire Western Ceramics glazes. 18" h x 12" w. *Photo: David Harder*

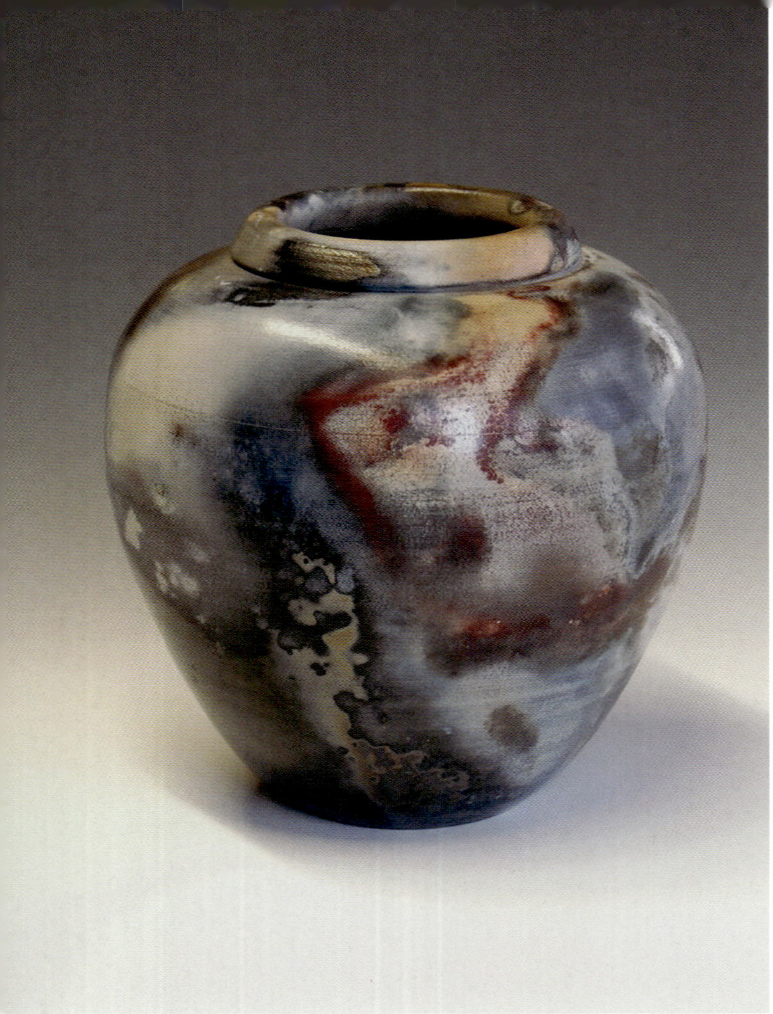

David Harder. *Bauer Jar.* Wheel-thrown using Aardvark-Sonora White clay, burnished, Terra Sigillata, buffed with a plastic grocery bag three times. 12" h x 10" w. *Photo: David Harder*

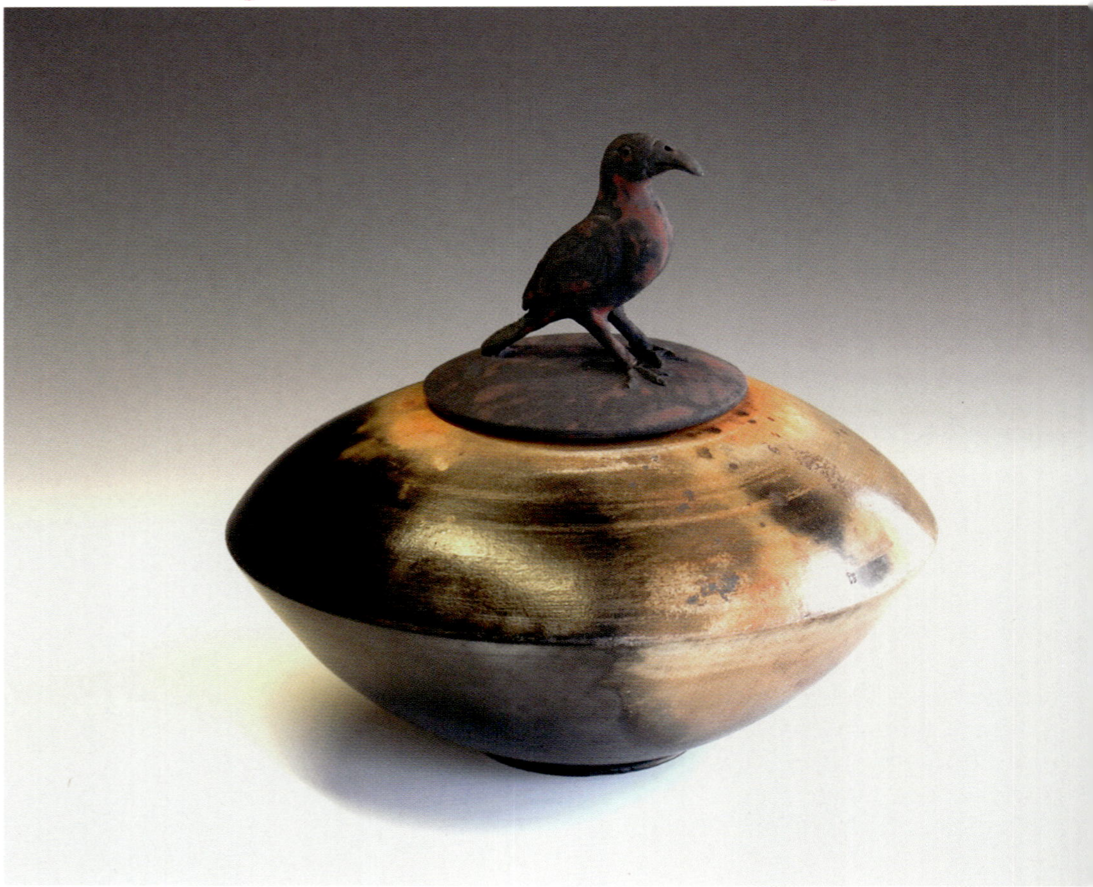

David Harder. *Seed Pot and Bird Lid*. Wheel-thrown using Clay Planet-Cinnamon clay, burnished, Red-Art Terra Sigillata, buffed with a plastic grocery bag three times. Lid—hand sculpted from paper clay and hand-painted with custom flat-black low-fire glaze. 6" h x 12" w. *Photo: David Harder*

VICKI HARDIN
(US)

Vicki Hardin, an established clay artist, has worked in clay since 1980, producing raku and pit fired pottery. She has shown throughout the Southwest, and extensively in Texas, where she maintains her studio, Clay Art Pottery, her home, and family. Her pottery is currently on view at Show of Hands Gallery in Philadelphia, Pennsylvania. She is an avid brush maker and teaches workshops throughout the USA. She recently completed a five day workshop at Penland School of Craft. Hardin gives back to the clay community by publishing the Clay Art Web Guide, an online ceramic portal.

Of her artwork Vicki Hardin says:

I have a fascination with hand building a form that is poised on the balance of crude elegance, which I have often explored to free myself from a production ethic. As a child, I was involved in the excavation of an American Indian village on my grandfather's farm. I find that I have an appreciation of the old, ancient works of a world that has moved on. Alternative firing processes are my interest. I have moved from a smoke firing process that utilized a considerable amount of sawdust, to utilizing pecan limbs and brush to stoke a pit for approximately two hours. Metallic salts are used within saggars to induce color and to create a smoke resist.

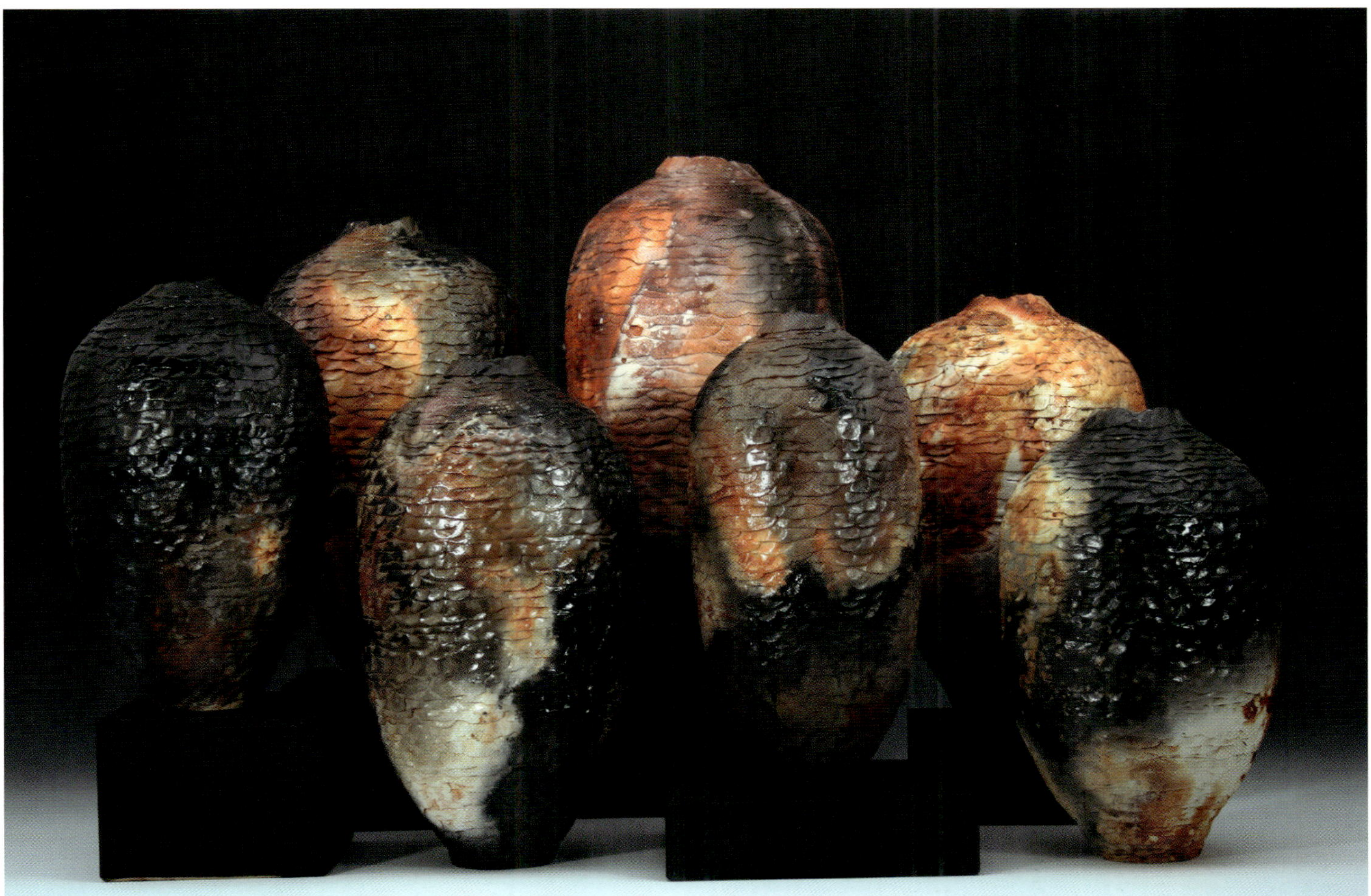

Vicki Hardin. *Untitled.* Handbuilt pit fired vessel. Pit fired in aluminum saggars with metallic salts. 10" h x 4" d & 14" h x 5" d. *Photo: Vicki Hardin*

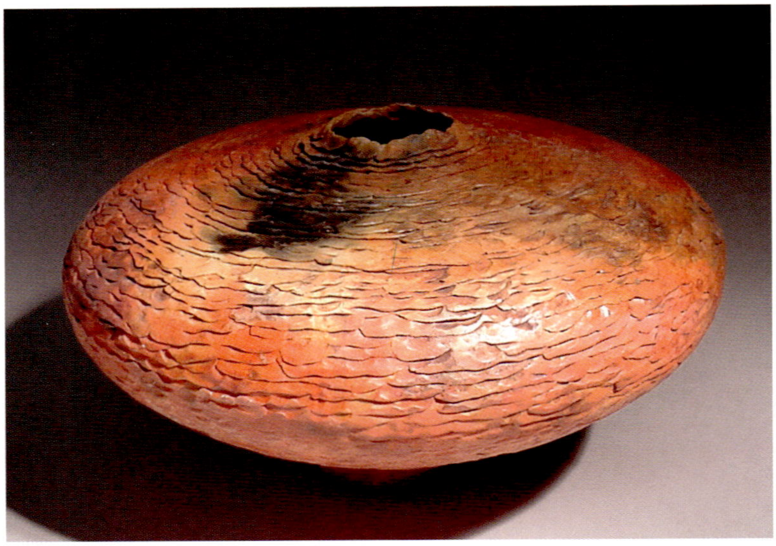

Vicki Hardin. *Untitled.* Handbuilt pit fired vessel. Pit fired in aluminum saggars with metallic salts. 14" h x 12" d. *Photo: Vicki Hardin*

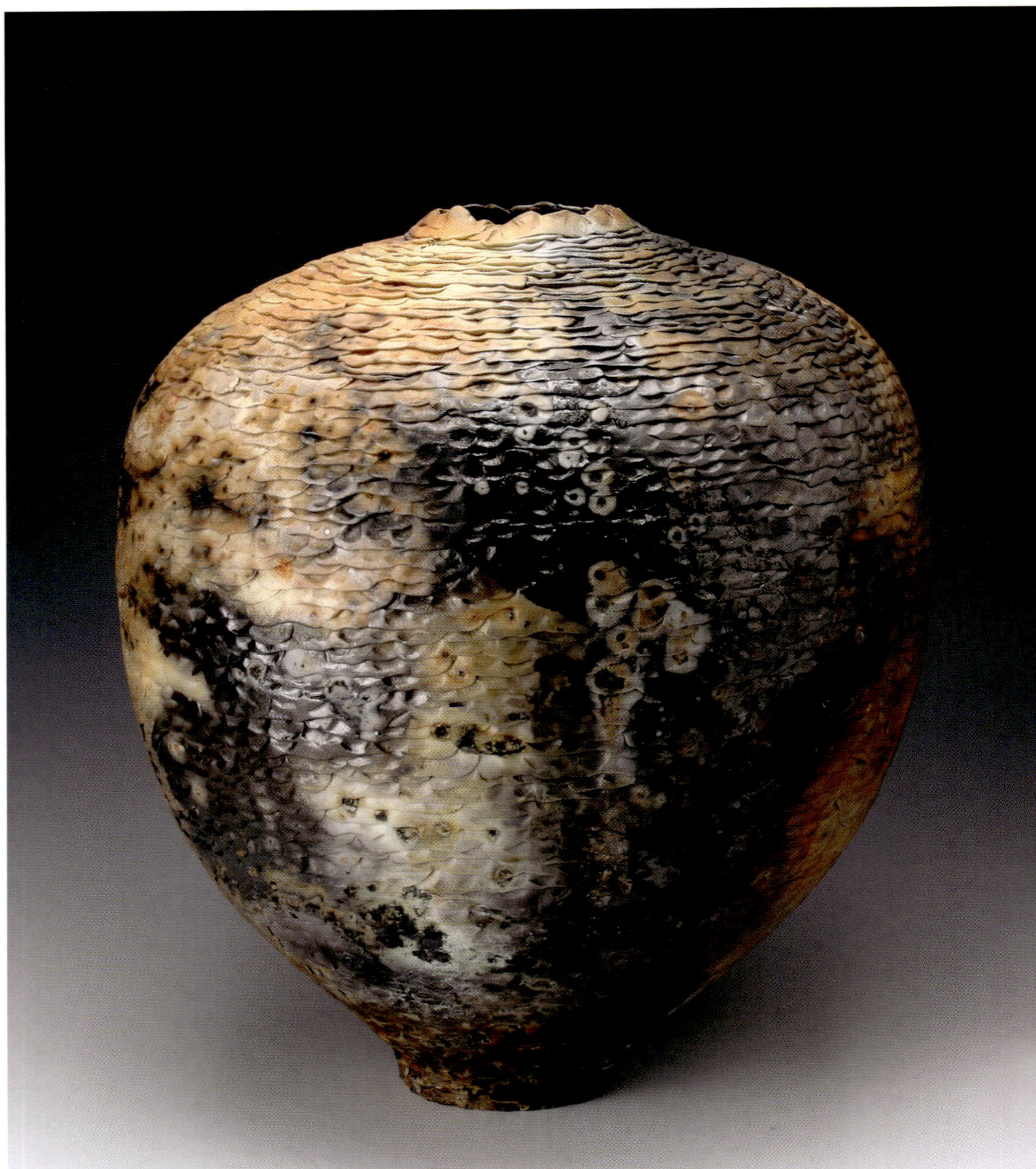

Vicki Hardin. *Untitled.* Handbuilt pit fired vessel. Pit fired in aluminum saggars with metallic salts. 24" h x 18" d. *Photo: Tony Ferguson*

BARRY HAYES
(AUSTRALIA)

Working in production potteries, where the aim is to replicate shape, glaze, and decoration, fertilized the seed of rebellion in Barry Hayes, which was planted after working in Papua New Guinea and seeing the traditional pots and making methods of Indigenous communities. After leaving production work, Hayes took up pit firing as a counter to that drone-like experience.

Hayes fires in a pit measuring 8' l x 5' w x 3' d, but explains that the only important dimension is depth: if a pit is too shallow it is difficult to maintain reduction, and pots are more susceptible to cooling dents. He prefers to use an iron-bearing clay as he finds this gives better color development. After burnishing, Hayes bisque fires his pots to 1650° F, which allows the clay to remain porous enough to absorb the fuming in the pit. A six-inch layer of sawdust is spread on the pit floor, and pieces are stacked: smaller pots around the perimeter to larger pots in the center. To attain black areas on a pot, fine sawdust is scattered beneath the pot, pinks are achieved by sprinkling copper carbonate, and rock salt is used for orange and yellow flashes. Hayes has found that experience is the judge of how much copper and salt to use: too much copper too close to the pots will cause damage, while too little and the vibrancy of color will not develop. Once the pit is stacked, timber is placed on top of the pots and corrugated iron is placed over the pit. A fire is lit at one end and encouraged to burn along the top of the timber by using the iron sheets as a shifting flue. Stoking is continued to avoid cold draughts on the hot pots, and to maintain a heavily reducing atmosphere. Hayes stops stoking when there is a good build up of coals, which usually takes about six hours. The pit is then sealed by packing dirt over the ends of the tin.

Barry Hayes. *Pot Form.* Handbuilt stoneware, burnished and pit fired. 9" h x 7" d. *Photo: Barry Hayes*

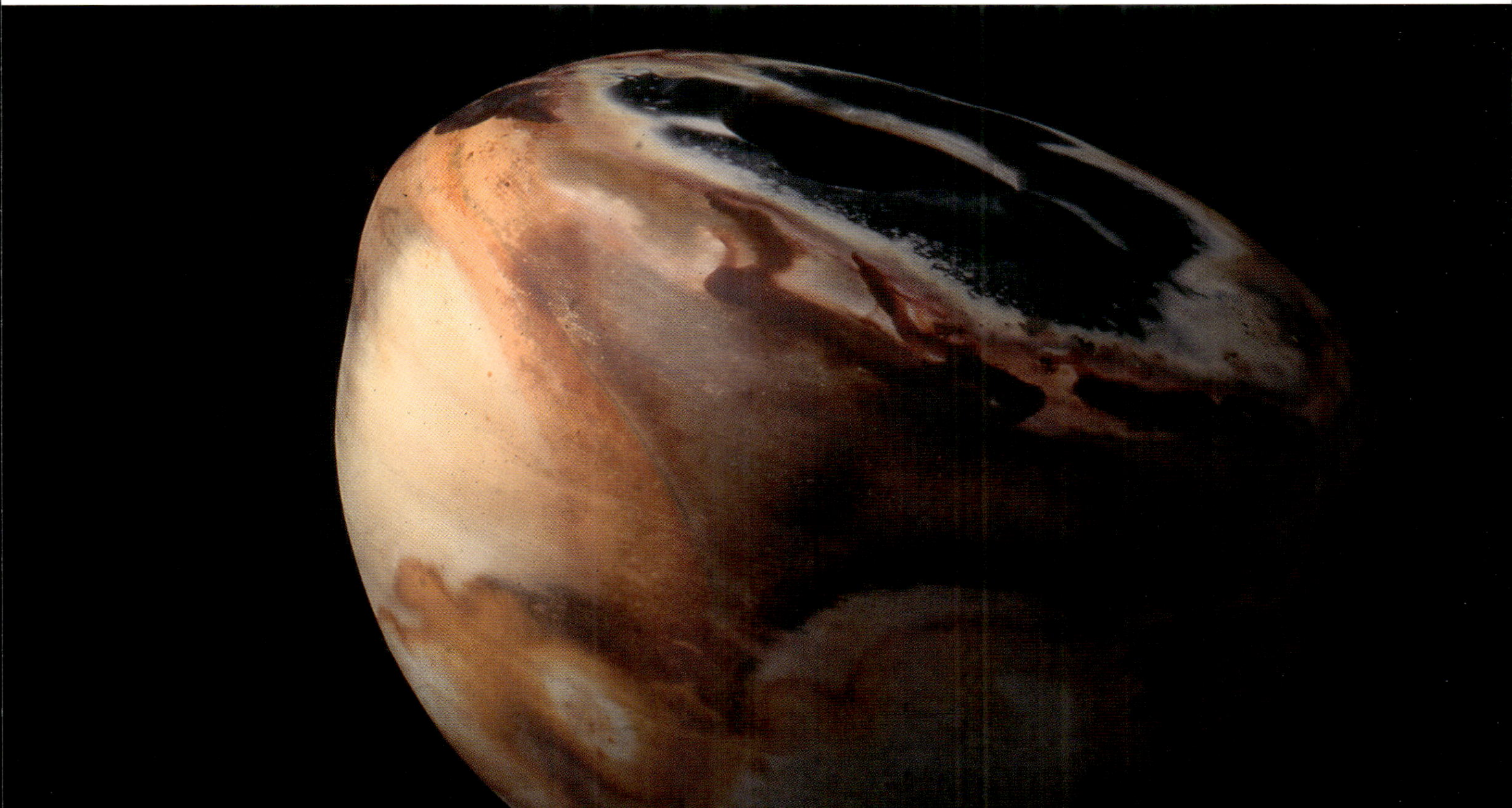

Hayes has exhibited widely in Australia in both solo and group shows, and his work is held in several Australian and international collections. He has received commissions in Australia, Spain, and Japan, and has been involved in the ceramics industry as a teacher and a committee member on several ceramic boards. Hayes has been featured in several periodicals and books, including the *Who's Who of Australian Visual Artists 1995* and *Smoke Fired Pottery* by Jane Perryman.

Of his art practice Barry Hayes says:

> Other factors also attracted me to this process. Firstly, the fire. The challenge of controlling the heat and atmosphere in the pit using only wood, air, experience, and gut instinct. Secondly, the serendipity of the process. After thousands of firings, unpacking the pit still manages to excite. Pit firing suits the forms that appeal to me. The simple, beautifully flowing forms of traditional Chinese pots are the perfect canvas. Surprisingly, and most importantly for a professional potter, the pit fired pieces sold, both here and overseas. Initially, galleries were unsure about the work because it was so different. Fortunately, the pots sold, so that I could continue pit firing as I feel the whole process really suits my personality.

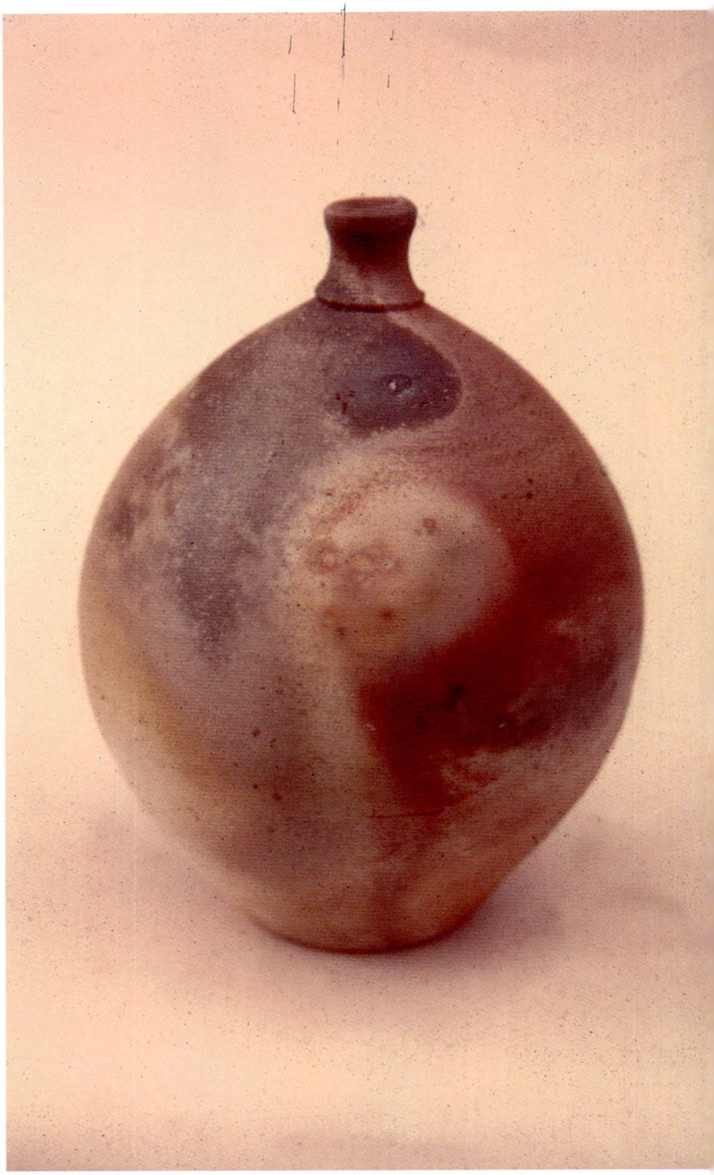

Barry Hayes. *Bottle Form.* Wheel-thrown stoneware, burnished and pit fired. 9" h x 7" d. *Photo: Barry Hayes*

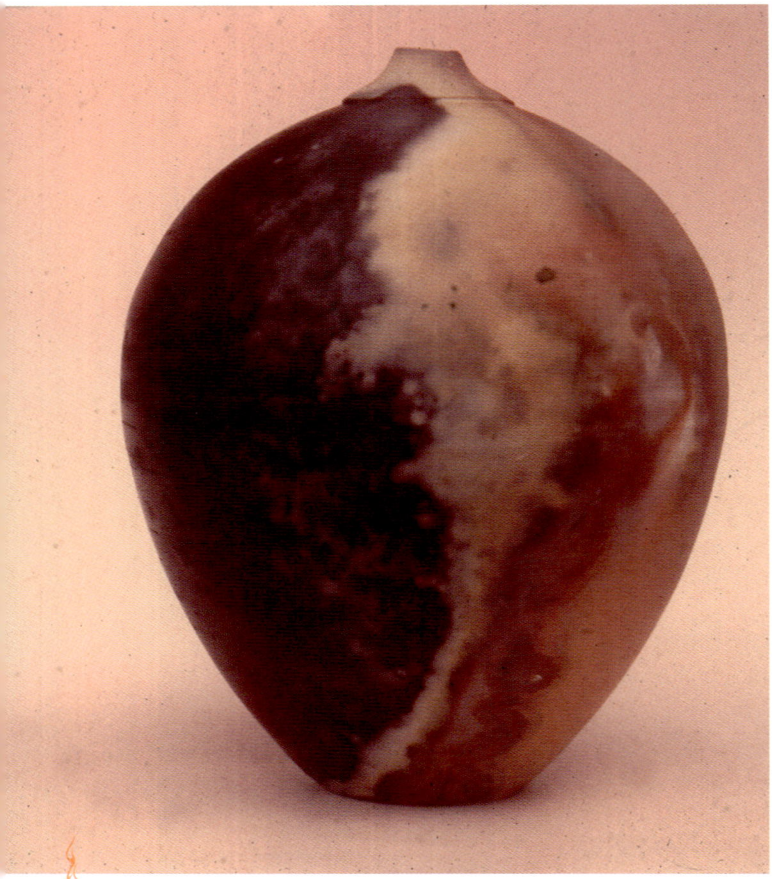

Barry Hayes. *Pot Form.* Wheel-thrown stoneware, burnished and pit fired. 12" h x 10" d. *Photo: Barry Hayes*

GUDRUN KAINZ
(AUSTRIA)

Gudrun Kainz was born in Vienna, Austria, and attended the School for Ceramics and Tilestove building from the young age of fourteen. Her early ceramic career was spent mainly making thrown functional wares using an electric kiln. These wares slowly morphed into the more abstract forms she currently makes after being influenced by several work and study stays in England with ceramicists such as Seth Cardew, Jane Perryman, and Richard Wilson (Perryman 2008, 78). Since 1998, she has been teaching primitive firing techniques to various community groups, and in 2007 and 2009 taught pit firing and other smoke firing techniques to orphans in a vocational training school in Uganda. During this time, she also taught students and tutors of the Art Department of the Makerere University of Kampala. Kainz has been regularly pit firing for thirteen years.

Kainz's forms are both wheel-thrown and handbuilt, though slab techniques have allowed her to increase the range of shapes she is able to produce. She uses a white or red Westerwalder stoneware clay, which she mixes with Limoges porcelain, though for larger pieces she uses a grogged clay. Leather-hard works are burnished with a stone and low bisque fired. Her pit measures 16' 4" l x 6' 7" w x 5' 3" d and is lined with stone due to the harsh winters in Austria (Ibid). A layer of sawdust is spread on the pit floor, which the bisque works are nestled into. Some works are wrapped in paper saggars containing oxides, salt, and organic materials, and other works are tied with copper wire and salted string. Between and around the works are sprinkled salt, copper carbonate, copper sulphate, red iron oxide, and organic materials such as banana peels, avocado skins, seaweed, straw, and grass. The pit is then filled with wood shavings and pieces of wood. Metal sheets cover the pit to control airflow. Firings usually take place at night and are unpacked on the third day (Ibid).

Kainz uses pit firing to give her work an extra dimension. The simple and discreet forms are the ideal partner for the dynamism of the flame, and her experimentation with small paper and slip saggars within the pit has resulted in a high level of color development of the pot surfaces. Gudrun does not use glaze, instead favoring unsealed open pieces, sometimes polished, sometimes not. She then combines the ceramic forms with other materials such as metal and wood, creating simple archaic forms that embody archetypes of movement and of existence.

Kainz believes that pit firing demands artists let go of their work completely; their influence on the result is minimal. One has to accept this challenge after the rather conscious and controlled part of forming the object.

Kainz enjoys working with open fire, viewing it as an element of human history.

Though Kainz's sculptures go far beyond the original intention of ceramics, she does not view her aesthetic understanding as a contradiction to ceramic traditions. Rather, her harmony of materials emphasize and extend the meaning of ceramic sculpture.

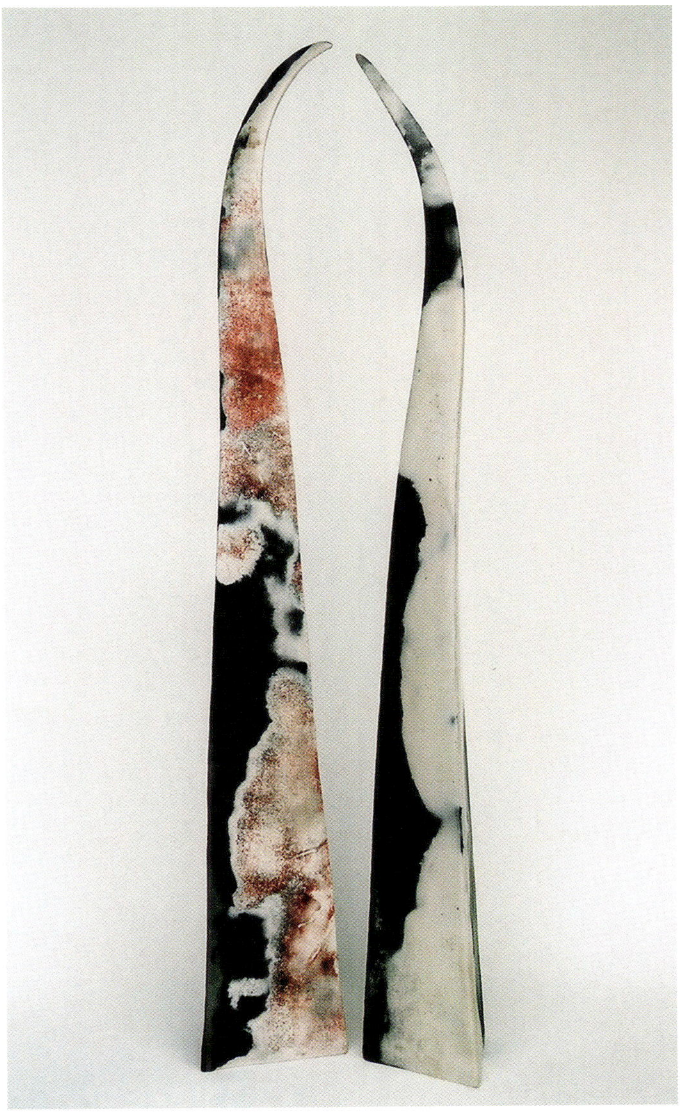

Gudrun Kainz. *Cones.* Slab built white stoneware clay (fine grogged), burnished, bisque fired to 1740° F, pit fired, no wax after pit firing. 26 ¾" h. *Photo: Artist*

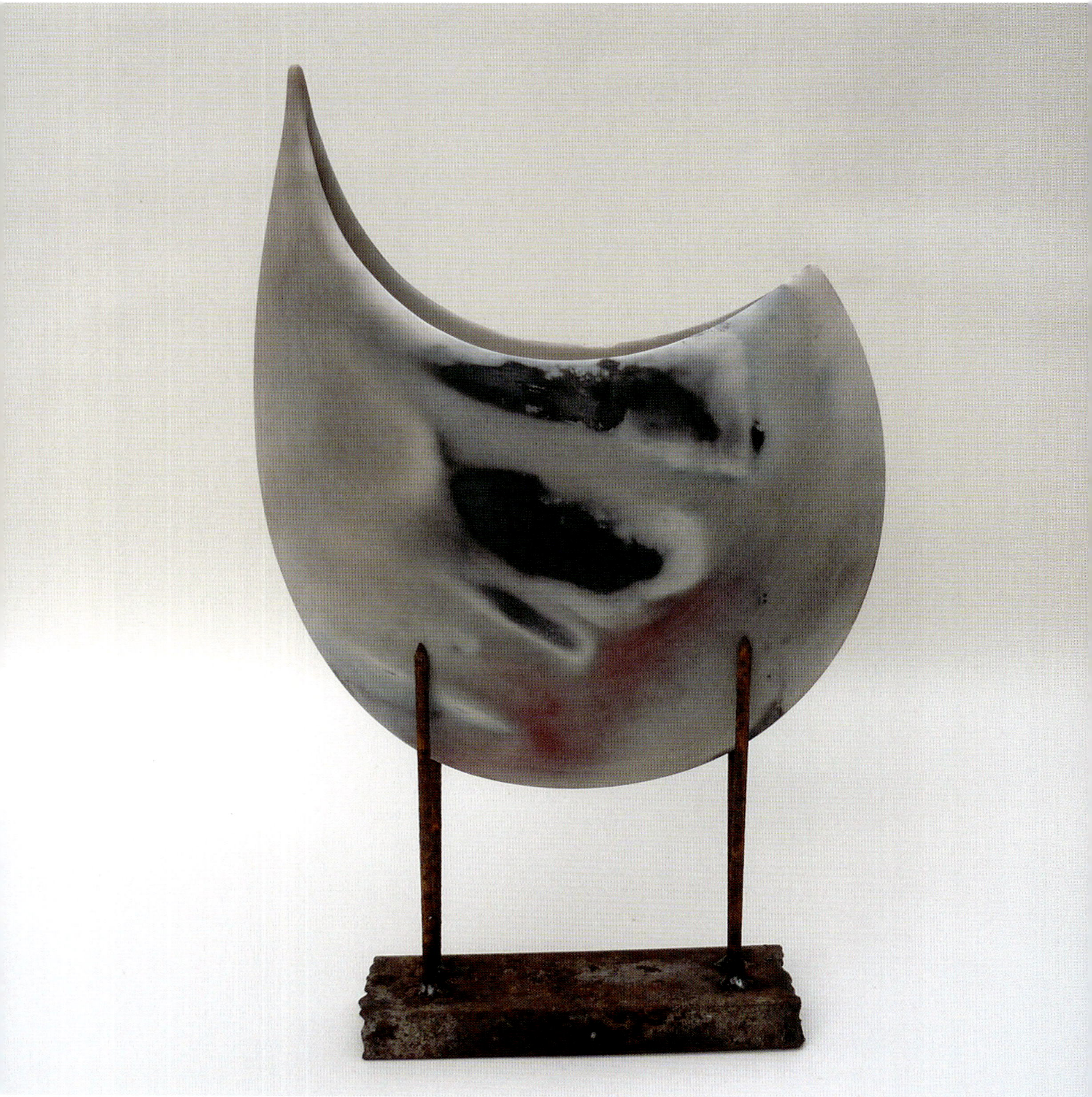

Gudrun Kainz. *Boat.* Slab built white stoneware clay (fine grogged), burnished, bisque fired to 1740° F, pit fired, no wax after pit firing. 15 ½" h x 10 ¾" w. *Photo: Artist*

MARY KAUN-ENGLISH
(UK)

Mary Kaun-English was born and raised in Southern California, where her passion for nature started as a young girl. Living in the then-undeveloped San Rafael foothills, she was allowed to wander in this desert landscape, filled with an inquisitive nature and fascinated by the wild world around her.

Kaun English moved to England in 1987, and returned to school, receiving a Bachelor of Arts (Hons) in the Practice and Theory in Fine Art and Design, from Wimbledon College of Art.

Since graduating from college, Kaun-English has exhibited her work widely and is represented by several galleries. Kaun-English currently works from her garden studio at her home situated next to the woods, which she finds to be a great source of organic inspiration for her practice. Her recent work includes hand built clay vessels and sculptures inspired by the ancient North American Indians, her biological ancestors. The second firing of these pieces occurs in a pit firing, exposing the porous clay to the smoke given off by the natural organic materials used to fuel the fire. The results are organic vessels and sculptures painted by nature herself.

Of her artwork Mary Kaun-English says:

> I have been inspired by nature and the natural spirit world for most of my life. I revere the strength as well as the forgiving life affirming forces of nature. My intention is to express these attributes reminiscent of the natural world in my work; using the varied forms and the emotive states of mind that I discover during my walks and frequent journeys, as inspiration.
>
> My thesis at university was based on Deluze's theory of "Becoming;" the idea that everything in our universe is in a constant state of transition. By using the low-tech method of pit firing, my ceramic sculptures have infused into their surface a picture of time's fluctuation. The traces of the ephemeral flame and smoke, from the fire, are left as an array of color flashes and smoke reductions through the body of the abstract forms.
>
> This process driven method is a collaboration with Mother Nature, involving an elaborate ceremony with its own set of rituals—each task having an important purpose. After weeks of preparation and a twenty-four hour firing, when I finally open the pit, a feeling of nervous excitement prevails; the sense of anticipation akin to finding artifacts on an archaeological dig. Then, examining each piece in detail to see what lay on the surface, trying to put together the puzzle of how each trace occurred.
>
> Using the method of pit firing, each piece I create is unique—nature never leaves the same mark twice. This method of working is not only fundamental to my artistic process but heavily addictive.

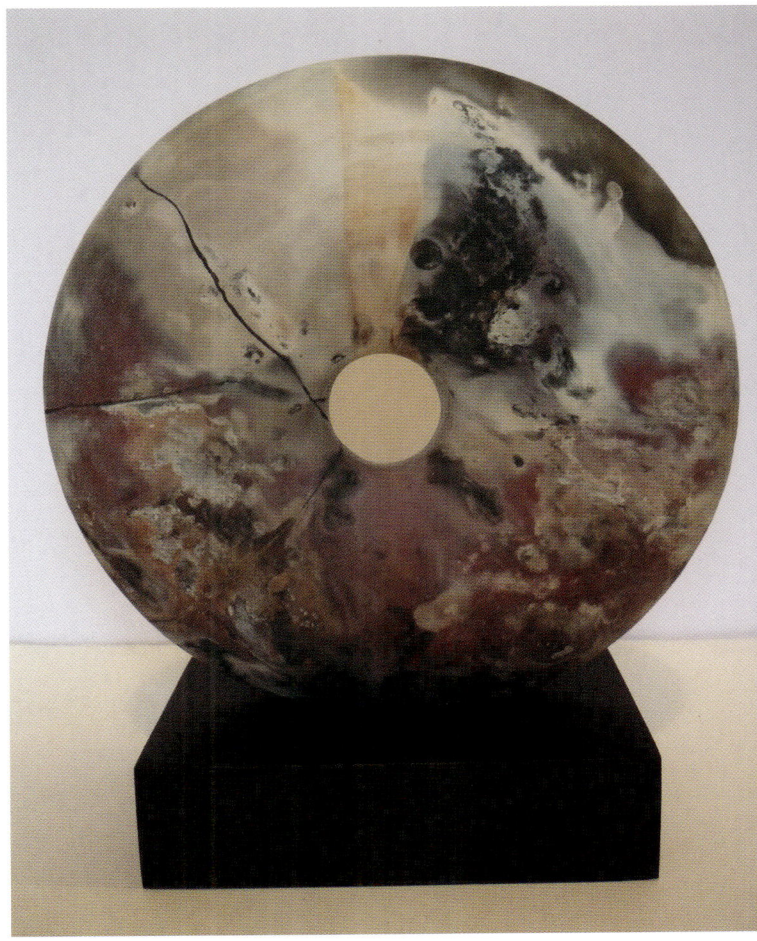

Mary Kaun English. *Dignity.* Handbuilt, pit fired ceramic. 13" x 11 ¾" x 2". *Photo: Mary Kaun English*

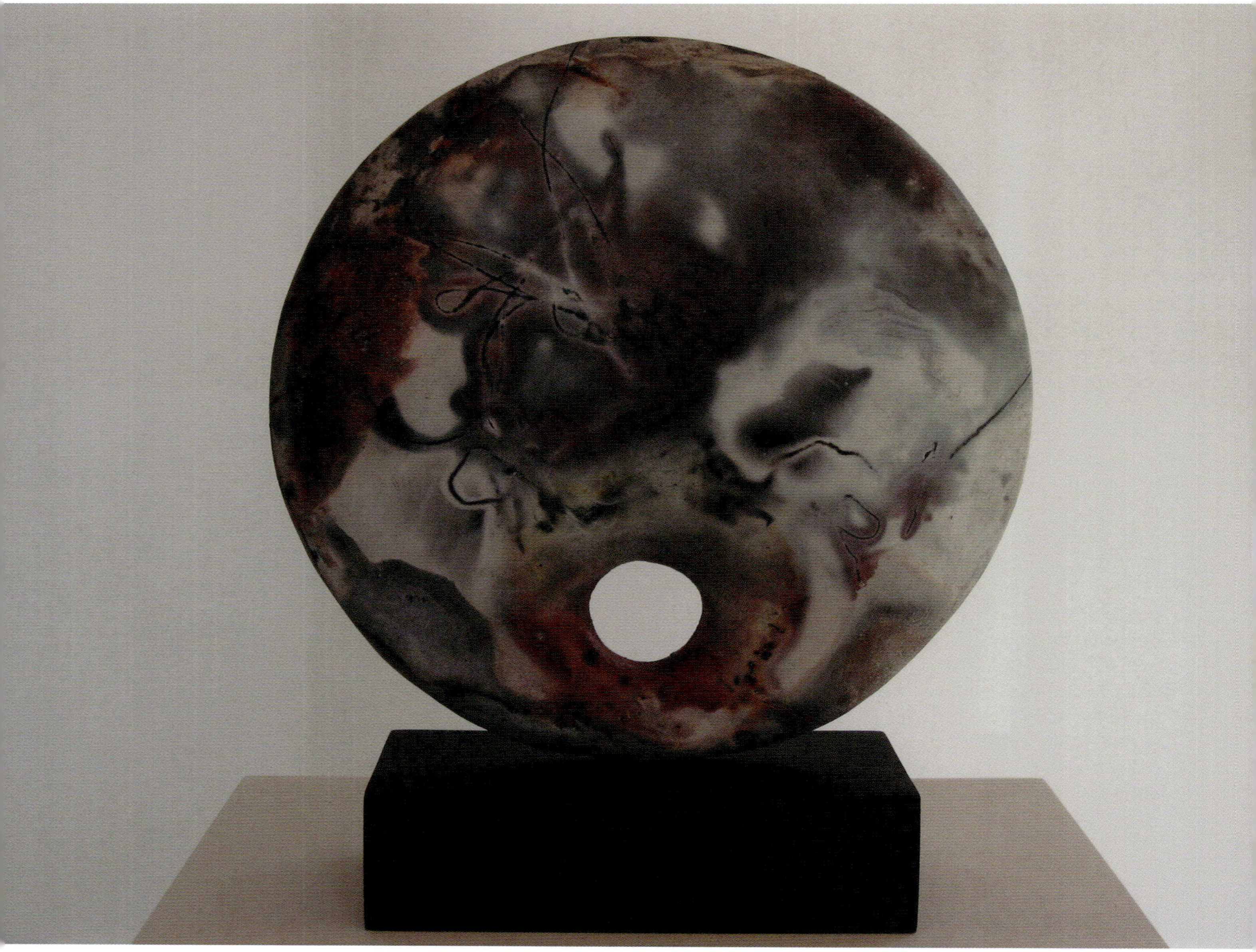

Mary Kaun-English. *Poised.* Handbuilt, pit fired ceramic. 13 ¾" x 11 ¾" x 2". *Photo: Mary Kaun English*

MARVIN KITSHAW
(UK)

Marvin Kitshaw is a potter based in Gateshead, England, UK. Trained as a 3-D designer, Marvin is a self taught professional potter who designs, makes, sells, teaches, and demonstrates through his business, Muddy Fingers Pottery. He specializes in functional stoneware, raku, and pit fired pottery, which he fires in local woodlands.

Of his work, Kitshaw says:

> I find pit firing a ceremony: making the ware, preparing their rite of fire by embalming the pots with cloths, metals, and vegetation, and anointing them with prepared liquids. Each pot different in its own way. Preparing the "kiln" or "pit" to contain the ware and fire. Layering the pottery packages and salty oxide parcels and timbers. Then the fire, intuition taking the front seat to control the burn. Clamming up the smoldering embers and waiting until the next day to open and excavate the burnt offerings the fire has painted. Cleaning, gently teasing, and revealing colors and buffing with waxes.
>
> The pit firing process is very different to the other techniques I use. It is natural, dirty, smelly, random, experimental, alchemic, intuitive, hot, cold, exhausting, exhilarating, and yields unexpected, surprising results. It is mud and fire. What else do you need?

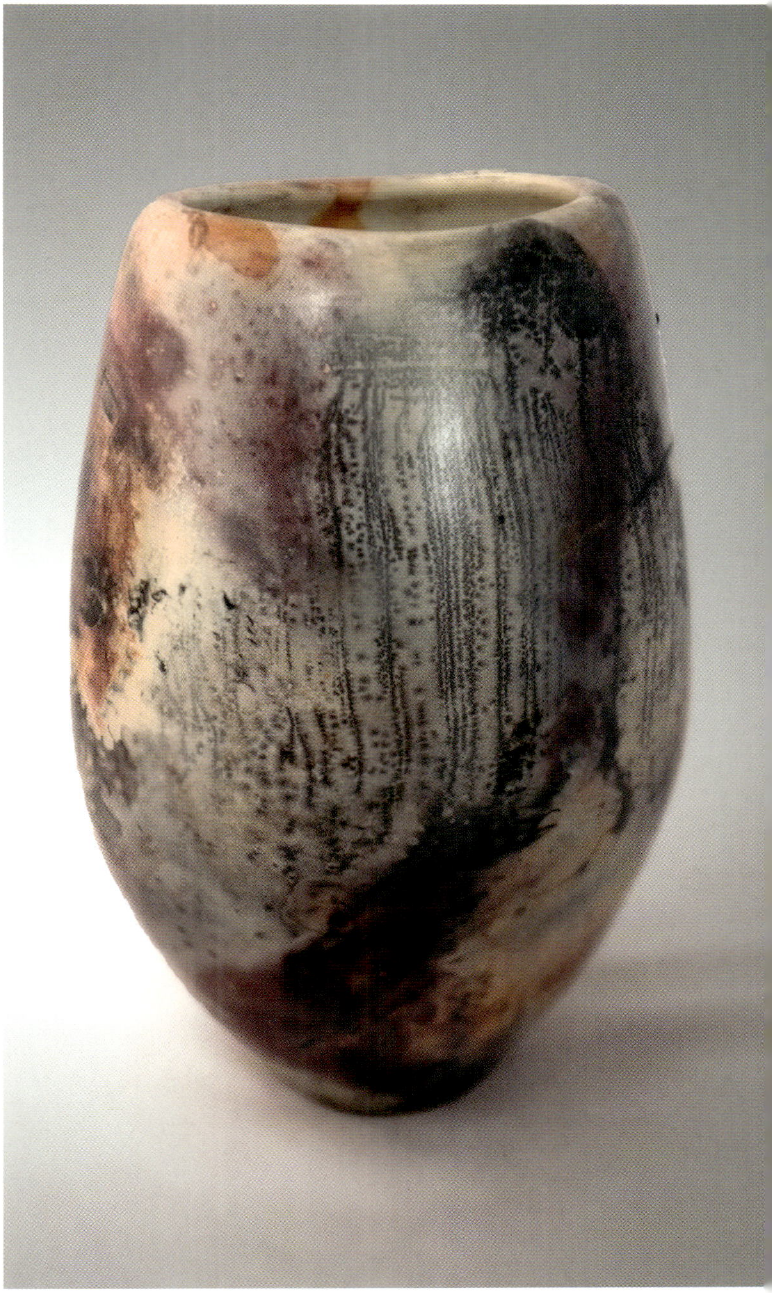

Marvin Kitshaw. *Water.* Wheel-thrown grogged stoneware, hand burnished with stone, plastic, and skin. 6" w x 9" h. *Photo: PollyB*

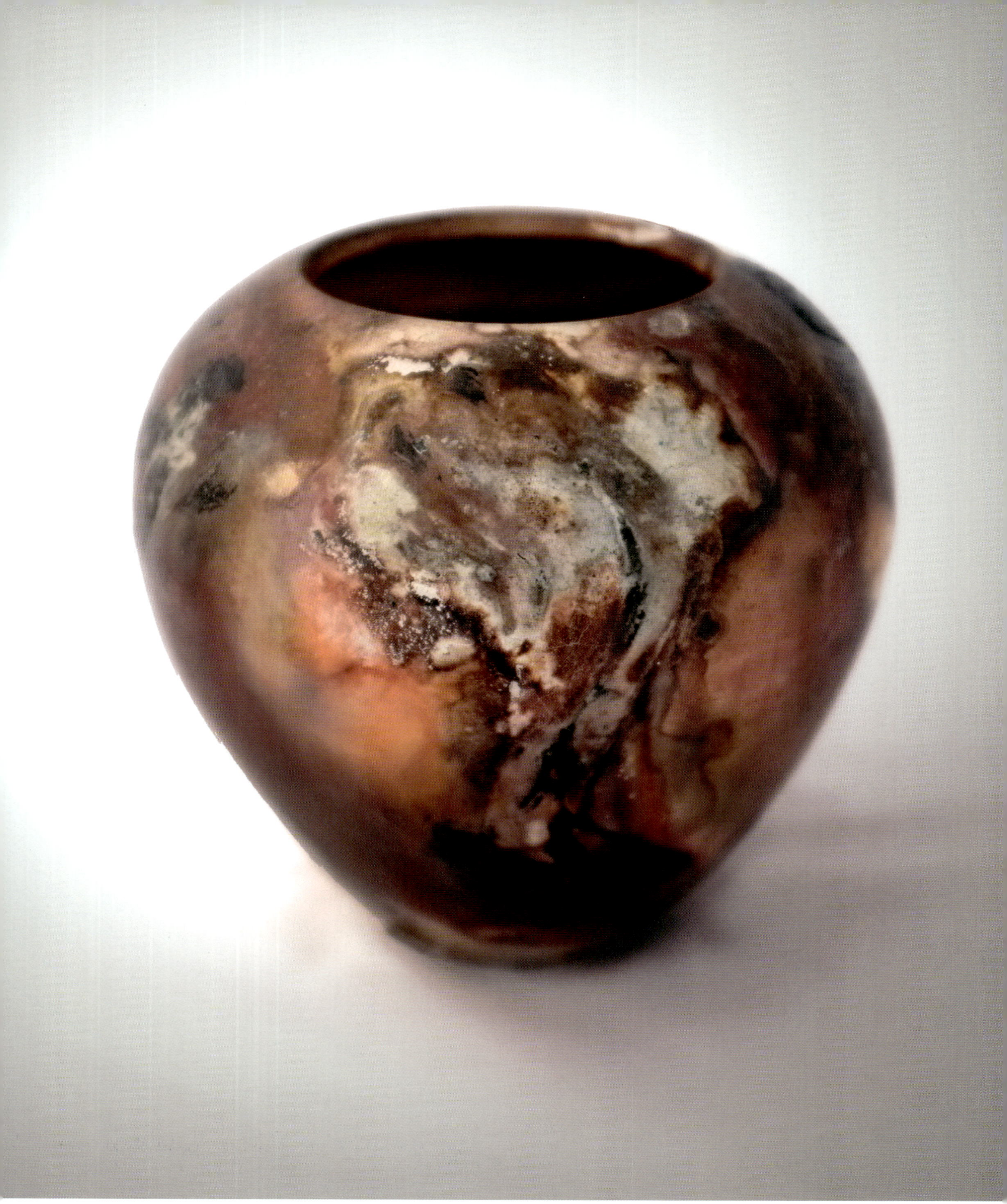

Marvin Kitshaw. *Planet Pot.* Wheel-thrown grogged stoneware, hand burnished with stone, plastic, and skin. 6 ¼" w x 6 ½" h. *Photo: PollyB*

GENEZ MALEBRANCHE
(US)

My work has often been an expression of two extremes. My technical illustration background gives me a desire to control the image, leading me to throw pots on the wheel with exacting precision. But my artistic background brings with it the desire to relinquish control and decorate with free-flowing energy and spontaneity, leaving the results to fate. Pit-firing lends itself well to this format. A well-constructed piece has a better chance of surviving the thermal shock that pit-firing can bring, but the results are wide-ranging and unpredictable. No matter how hard I try to duplicate previous results, each piece is unique, like a fingerprint. And the results can be very dramatic.

Originally from Williamsburg, I have lived in the Tidewater area of Virginia all my life. Although I have been an artist since I was very young, I only discovered the potter's wheel in 1995. Since then, clay has been my passion. I took my first pottery class at the local Parks and Recreation facility, but have since taken classes and workshops with many renown pottery artists, including Joyce Michaud, Steve Branfman, Ramon Camarillo, Cynthia Bringle, Don Ellis, and Randy Broadnax.

I have participated in many shows in southeastern and central Virginia, including The Hand Workshop Craft and Design Show in Richmond, An Occasion For The Arts in Williamsburg, and the Boardwalk Art Show in Virginia Beach, where one of my pieces was selected for judging in 2007.

In January 2006, I was invited to participate in the Thou Shalt Not exhibition at the Visual Art Center in Portsmouth, Virginia, which included nine other well-known artists. In the Gallery section of the September/October 2006 issue of *Clay Times* you will find one of my pieces from this exhibit. Another of my pieces was selected to be included in the book *500 RAKU* published by Lark Books.

I am a member of The Ceramic Designers Association of Tidewater, Virginia, and I currently design their newsletters. I have taught workshops for them and also for the Museum of Contemporary Art in Virginia Beach, Virginia.

I display my work at several galleries throughout the Tidewater area and in North Carolina.

I live on the Virginia side of Knotts Island with my wonderful husband and two beautiful children.

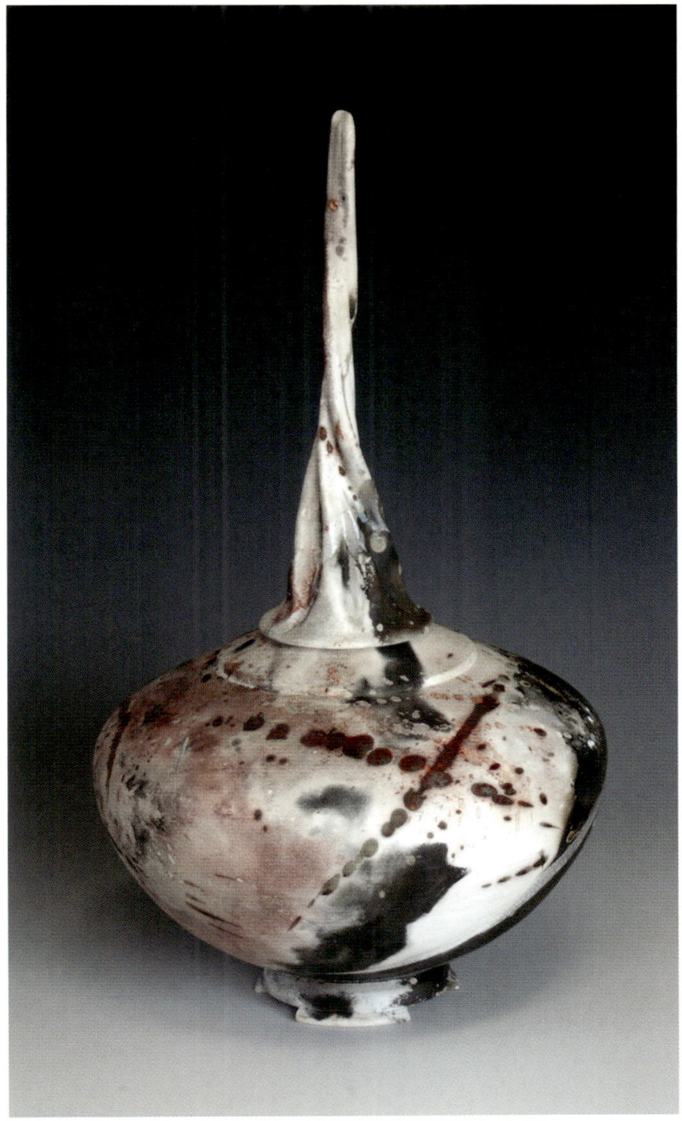

Genez Malebranche. *X Marks the Spot.* Porcelain, wheel-thrown, carved foot, altered lid, splashed with raku glaze, pit fired. 17 ½" h x 10". *Photo: Genez Malebranche*

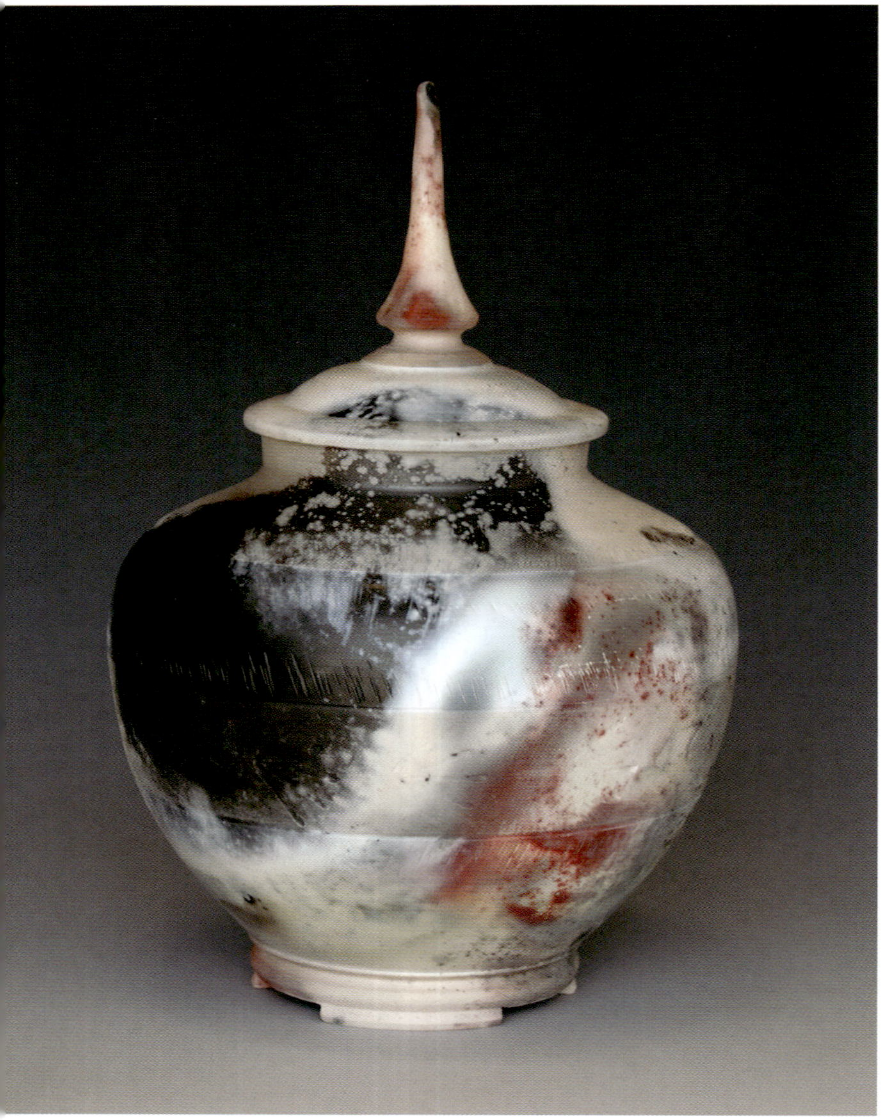

Genez Malebranche. *Red Sky At Night.* Stoneware, wheel-thrown, chattered, stamped, and carved, brushed with terra sigillata, pit fired. 9 ¾" tall x 6 ½" diameter. *Photo: Genez Malebranche*

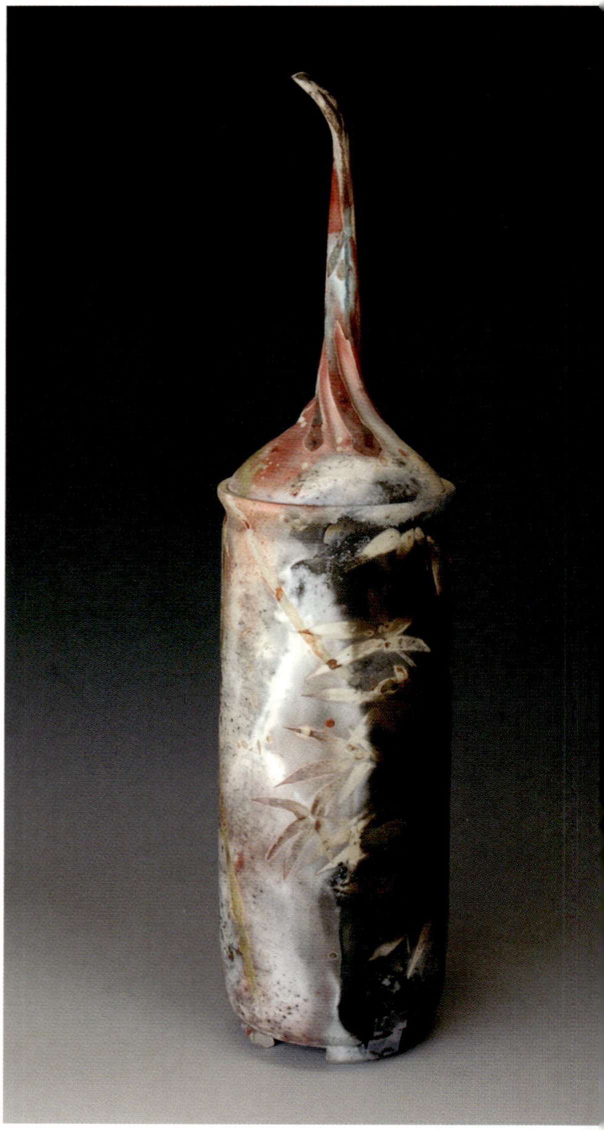

Genez Malebranche. *Burnt Bamboo.* Porcelain, wheel-thrown, carved foot, altered lid, brushed with raku glaze, pit fired. 19 ½" h x 4 ½" diameter. *Photo: Genez Malebranche*

ALEX MANDLI
(US)

My work encompasses over thirty years of experience and experimentation. From clay, water, and fire, I aspire to create simple forms influenced by an appreciation of Asian ceramic forms in combination with the surface decoration in the firing technique that replicates the natural color of stone.

The clay is molded by my hands and then given over to the fire. The end result is a finished object that is stronger and more beautiful than that which could have been done by my hands alone.

Throughout my life as a potter, my work ethic has been inspired by the fact that my time on earth is limited. Each human being wants to leave behind a legacy of their time on earth. For any artist, that legacy is the body of their work. The reality of my own mortality keeps me focused on my creative endeavors. The expression of my life's work can be seen in the trail of work I leave behind.

Since I began working in clay, my goal has been to make beautiful vessel forms and adorn those forms with surface decoration. Many times that decoration follows a theme or series. My current work involves the very basics of making pots, namely, the use of clay, water, and fire.

For years, I've studied and enjoyed the pots of many ancient cultures. Through this study, I became interested in their firing techniques, specifically their firing of work in a pit. I became attracted to the random action of fire on clay, or what has been termed "the gift of the flame." The excitement of this firing technique is the pleasant or sometimes unpleasant surprise discovered once the pit fire has cooled. Once the fire is cool, all that remains are the random marks made by the fire and ash.

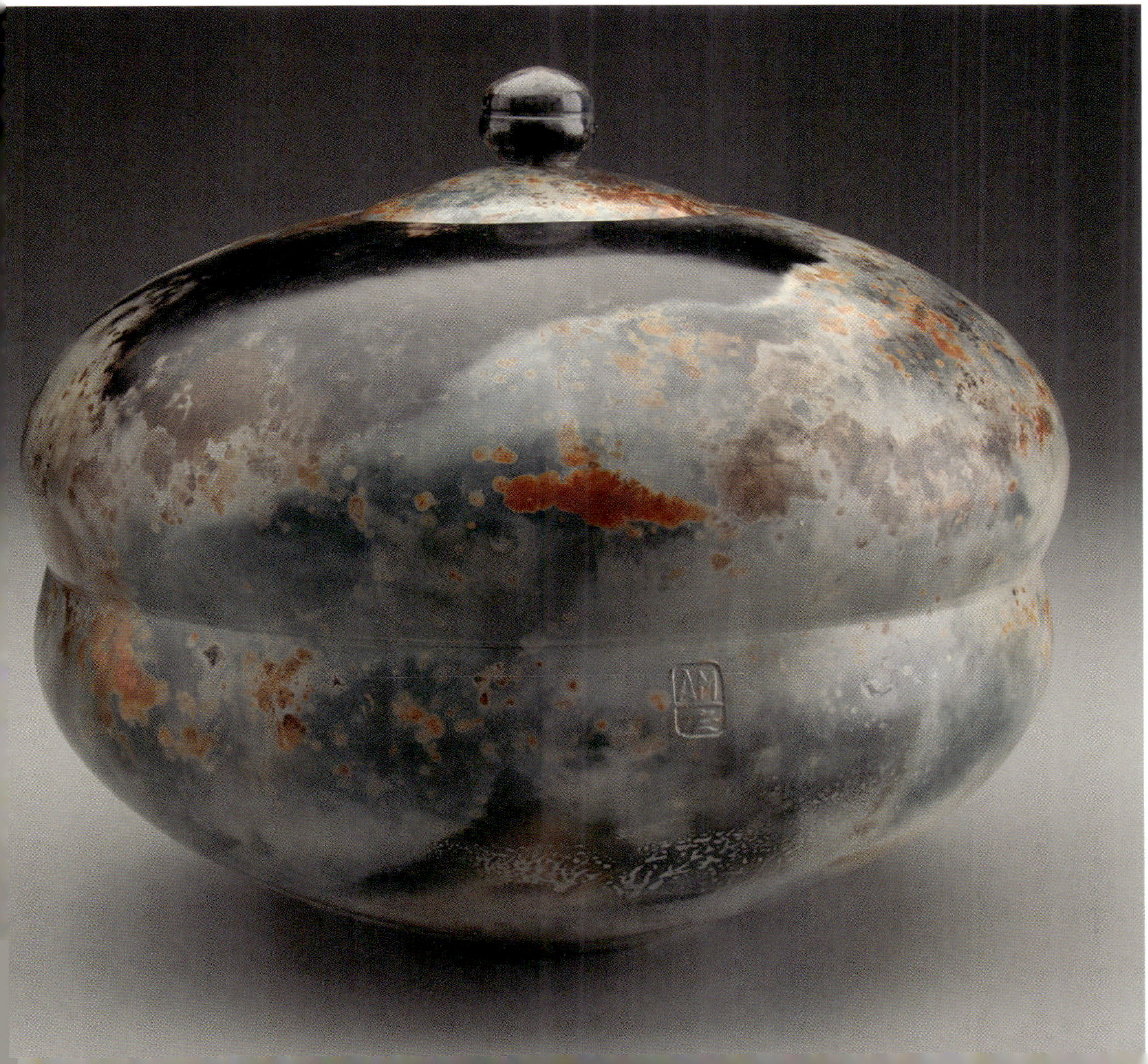

Alex Mandli. *Belly Button Covered Jar.* Wheel-thrown and burnished, terra sigillata. 14" h x 11" w. *Photo: William Lemke*

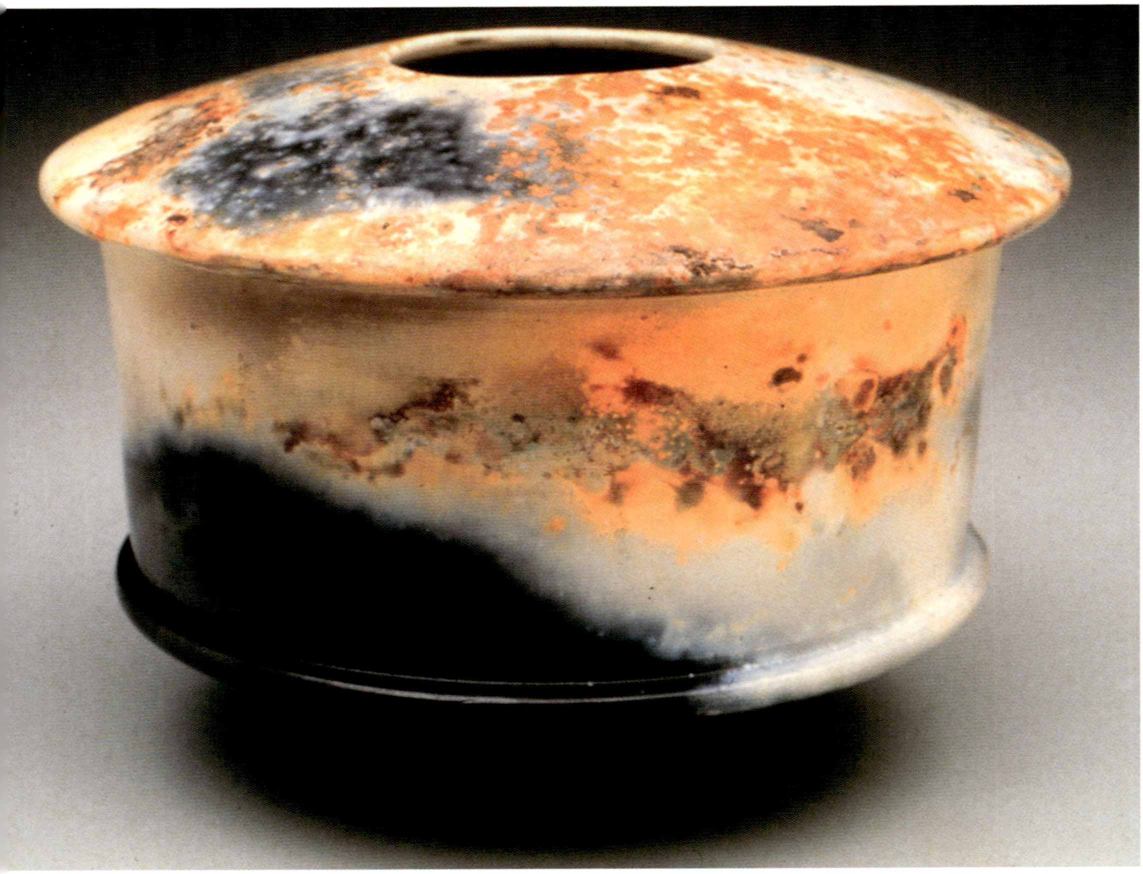

Alex Mandli. *Flat Top Jar.* Wheel-thrown and burnished, terra sigillata. 7" h x 9" w. *Photo: William Lemke*

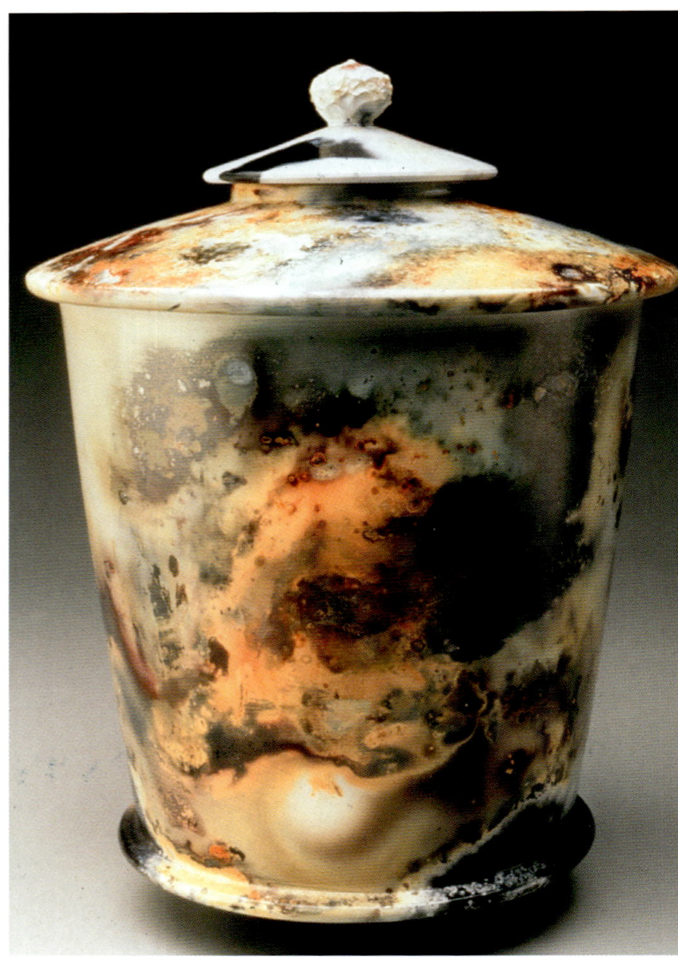

Alex Mandli. *Covered Jar.* Wheel-thrown and burnished, terra sigillata. 10" h x 8" w. *Photo: William Lemke*

CYNTHIA MCDOWELL
(US)

Cynthia McDowell attended Colorado State University and graduated with a BFA in 1998. She has been working full time in ceramics for about eight years in New Mexico. McDowell's focus is on pit firing, and also wood firing. She has built a pit kiln and wood kiln on her property, near the studio. McDowell's Earthfired Pottery Ranch thrives on creativity and hard work. She is happy to have other potters involved with her firings. While McDowell finds it convenient and productive to work from home, she also enjoys staying involved with the local college ceramics program, regional wood fire community, and attending workshops. These interactions keep her motivated, learning, and well socialized.

McDowell's pottery is sold through various outlets, including juried art shows, handmade markets, and local galleries. She also has an online presence. In the future McDowell hopes to reach out to a broader audience through teaching workshops from her studio.

Of her art practice, McDowell says:

> For many years, New Mexico has enchanted me with its high desert landscape, contrasting light, and rich history. I've spent countless afternoons hiking and horseback riding canyons, pondering petroglyphs, and wandering ancient ruins. Although my family does not stem from these roots, I have a deep respect for the ancestral Pueblan culture, which used to thrive in this rugged landscape. Sometimes, I find a coiled or painted pottery shard to temporarily hold and appreciate. It's an immediate connection to the past through something handmade. Their pottery was crucial to survival, yet function—"craftspersonship" and artistic choices are definitely present. The 1000 year old precise forms, clay bodies, and painted symbolic patterns are remarkable. This is where the inspiration for my pit fired work is drawn.
>
> The pit firing process is sacred to me and serves as a link to ancient techniques and people. My "modern" interpretation creates a lot of color and line, adding mystique to the decorative forms. I use high fire clay bodies, which vary in color and add lots of "magic dust" to the pit. My pots are wheel-thrown, burnished, and bisque fired. Then, I carefully place the pottery in a brick lined pit containing sawdust and other combustibles. I've had success adding materials such as hay, borax, copper, iron oxide, ferric chloride, soda, salt, and even coffee. I ignite the fire with local cedar and pine wood. Slowly, I bring the fire to the desired temperature, building a solid bed of coals, then cover and cool for a few days. Upon unloading the kiln, you never know what color pattern or design will occur. There is an imaginative, whimsical feel to the overall surface of the pot that connects with the viewer.
>
> Throughout my clay journey, I hope to pass on well made, intriguing pottery techniques and great friendship.

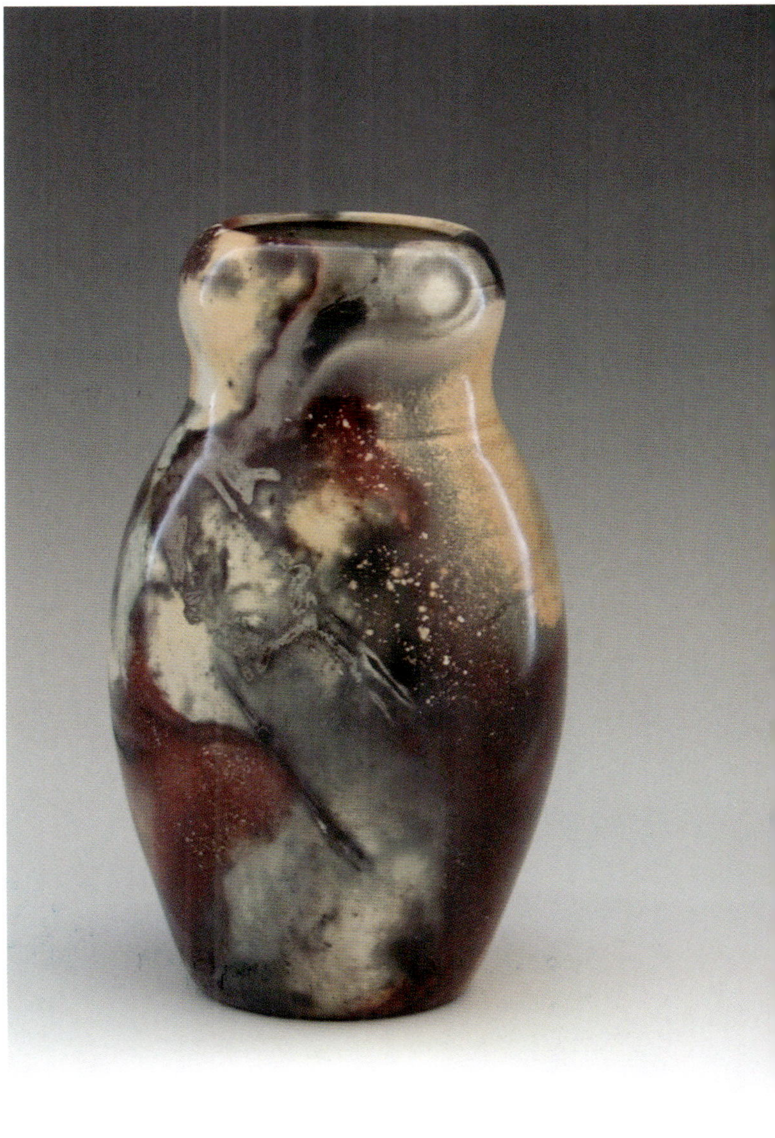

Cynthia McDowell. *Pit Fired Pot.* Wheel-thrown and burnished. 7" h x 4" w. *Photo: Cynthia McDowell*

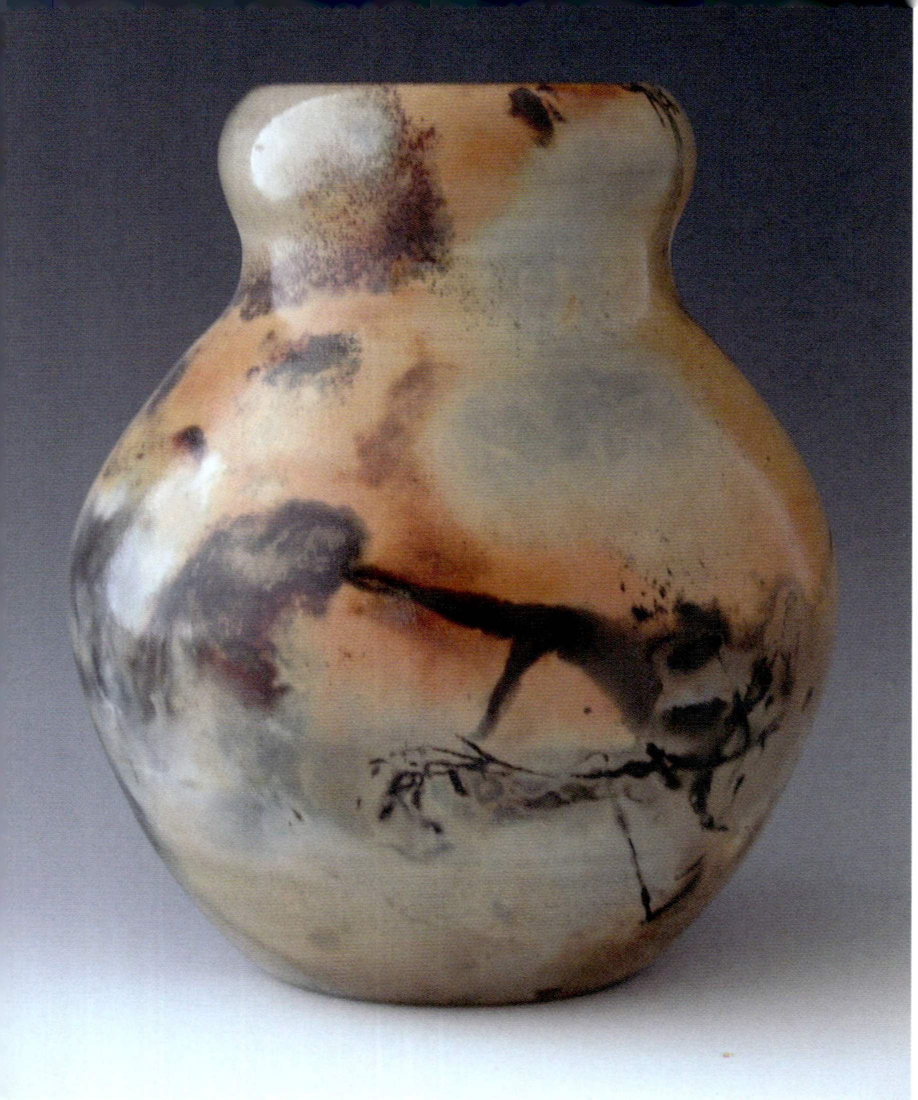

Cynthia McDowell. *Metamorphosis.* Wheel-thrown and burnished. 11" h x 8" w. *Photo: Cynthia McDowell*

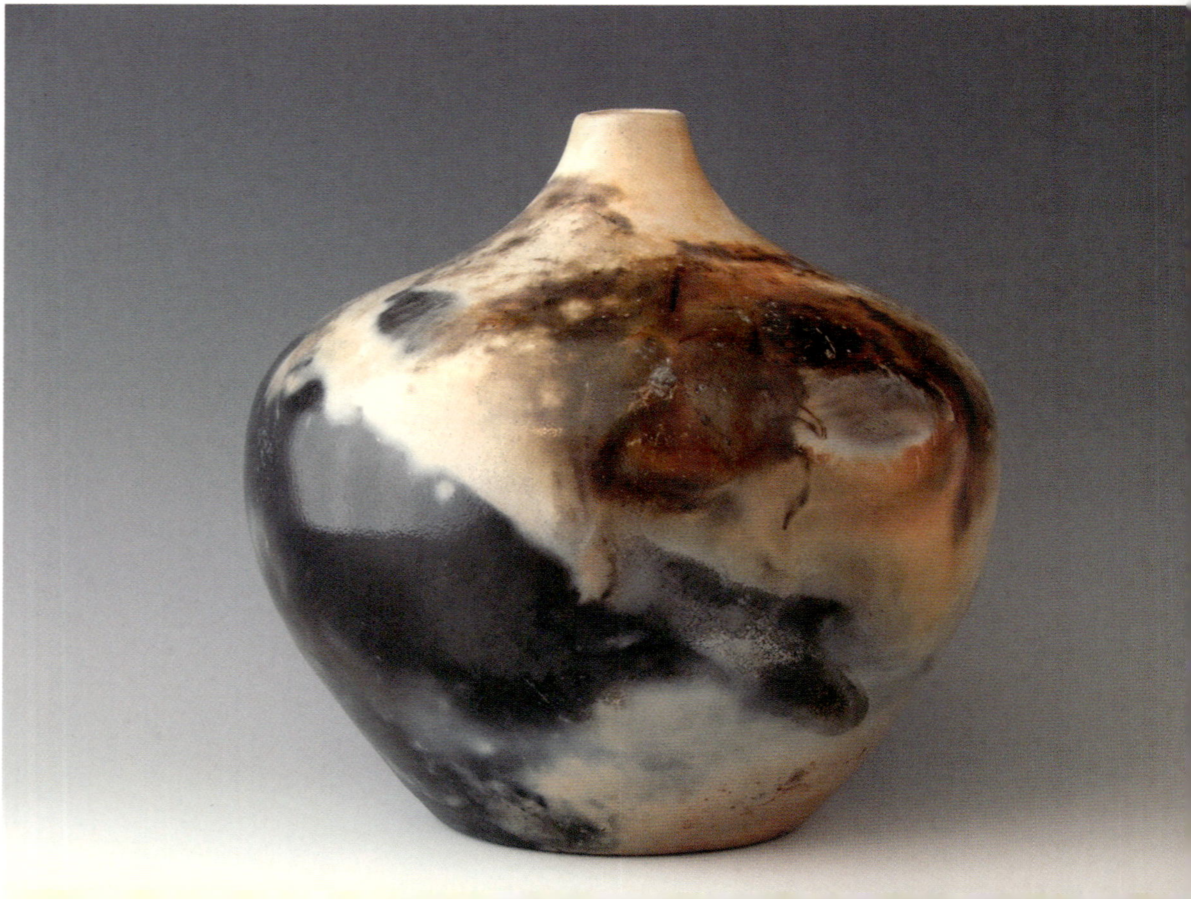

Cynthia McDowell. Venus Rising. Wheel-thrown and burnished. 9" h x 9" w. *Photo: Cynthia McDowell*

CAROL MOLLY PRIER

(US)

Carol Molly Prier's early ceramic influence was Inuit sculpture, and the discovery that she could make clay look and feel like polished bone after burnishing led her journey toward pit firing. Her first major work was an installation entitled *Demarking Space* assembled at Mills College in Oakland, California, in 1977 as part of her M.F.A. project. The installation consisted of eleven poles, nine feet high, burnished, and installed over a wooden armature. After working with large installations during her education, Prier wanted to work on a more intimate scale and began making small *Mystery Sticks*: curved bone-like pieces with tiny beads of clay inside that made a noise when lifted and moved, revealing the inner space of the object. During this time, Prier used what she called the "cheater" method: pre-heating her work before consigning it to the pit, which was already prepared with hot coals.

During this time, Prier also experimented with burnishing terra sigillata, but found it too difficult to work with. It would crack and flake. Now she burnishes the clay surface four times, and lets the clay dry between each burnish. Most of Prier's works were sculptural until 1983, when she began to make burnished functional forms influenced by the Pueblo pottery she had seen in books. She did not, however, decorate the forms in the Pueblo style, but instead wanted the surfaces to show the marks of the fire (Prier 2001, 31-3).

Carol Molly Prier says of her art practice:

> Whether making a cup or bowl for everyday use or a one-of-a-kind burnished vessel, I am guided by three concepts: simplicity of form, evocation of memory, and sense of ritual. The methods, the forms, and the rituals of the cultures creating before me inform my work.

Carol Molly Prier. *Tall Red.* Wheel-thrown and altered. 7 ½" x 7" x 4 ½" h. *Photo: Carol Molly Prier*

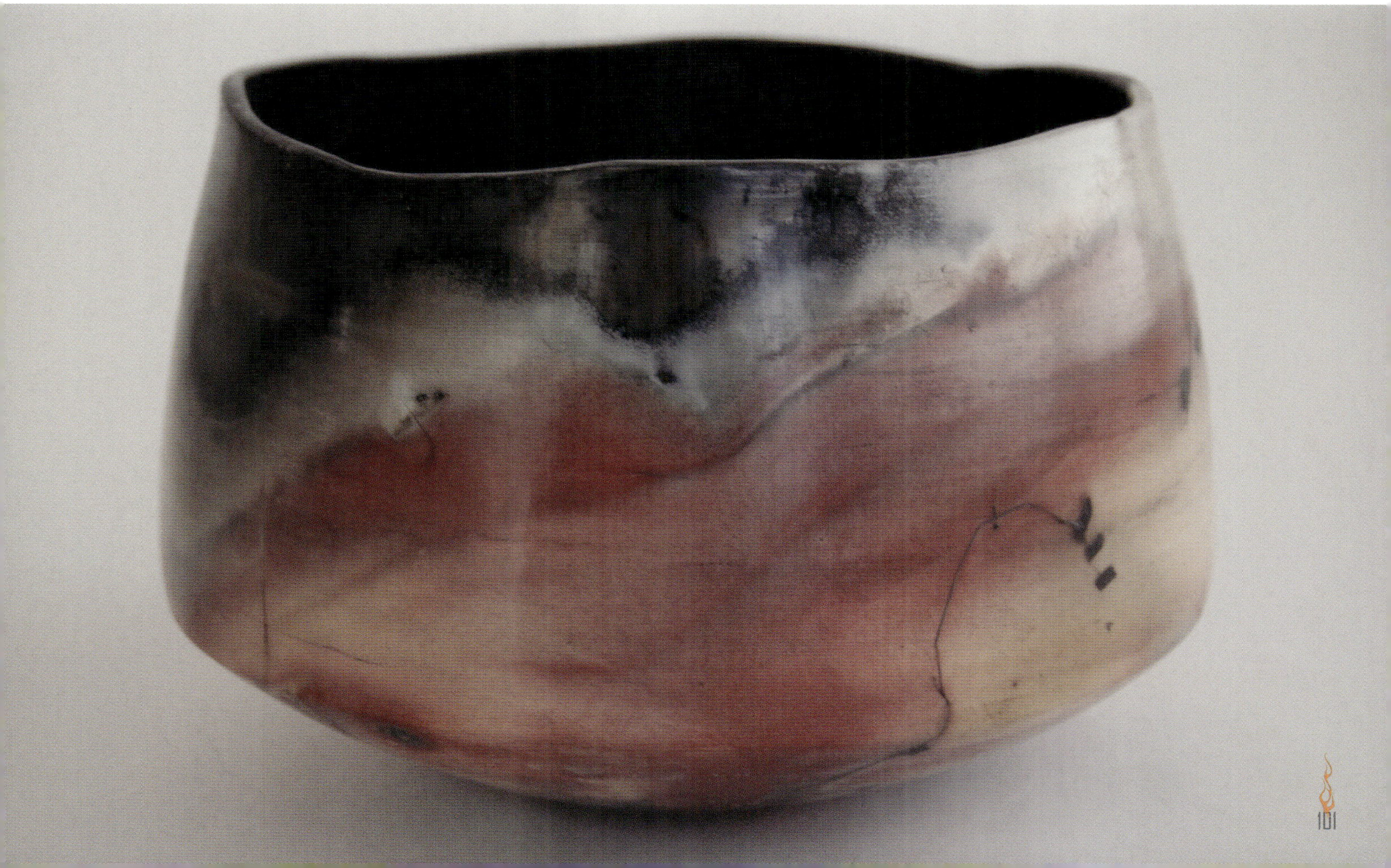

In 1974, while making sculptural pieces, I began teaching myself the processes of burnishing and pit firing. I had studied old Inuit sculpture made of polished bone or soapstone. Here, in clay, through burnishing and pit firing, which I learned by reading about Pueblo pottery processes, was a way to create a similar patina. I was also drawn to Pueblo pottery processes because of the lack of technology and equipment needed for creating these beautiful pieces. I feel how something is made comes through the final piece. I decided not to paint designs on my pieces, as Pueblo Native Americans do, but rather, to explore the markings of the fire, the "cloud marks" that the Native Americans try to avoid, as these obliterate their painted designs. I was working "backwards." But in this "backward" way, I have learned how the fire can mark the pieces in ways I could never achieve with glaze or how the fire can create a surface of soft, complete blackness.

Since about 1985 I have been making vessels reminiscent of Pueblo storage jars and now, more recently, round-bottomed Pueblo bowls. The forms of the old vessels are primal, strong, and beautiful. In my functional pieces, I strive to maintain the same simplicity and strength of form. I hope, as people experience my work, whether a cup, a bowl, or a burnished vessel, that something slows, then quiets, and perhaps an echo of memory comes.

In 2010 Carol had images of her work published in *Ceramics Today* (Schiffer Books).

Carol Molly Prier. *Untitled* detail. *Photo: Carol Molly Prier*

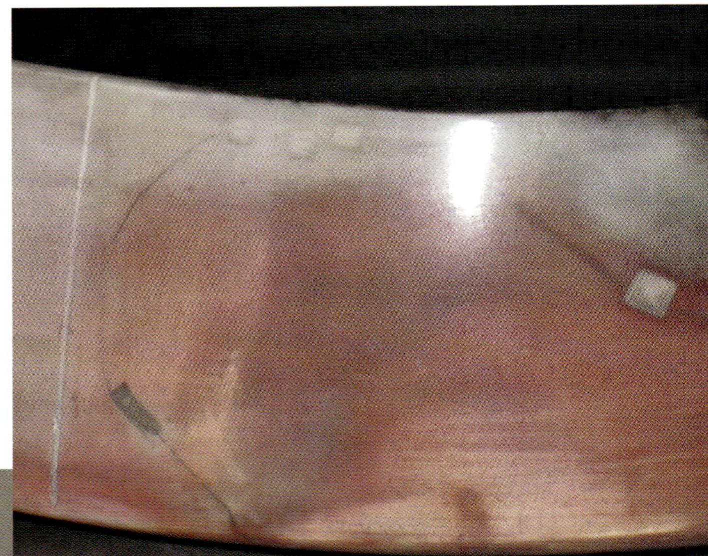

Carol Molly Prier. *Untitled.* Wheel-thrown and altered 8 ¼" x 8" x 4" h. *Photo: Carol Molly Prier*

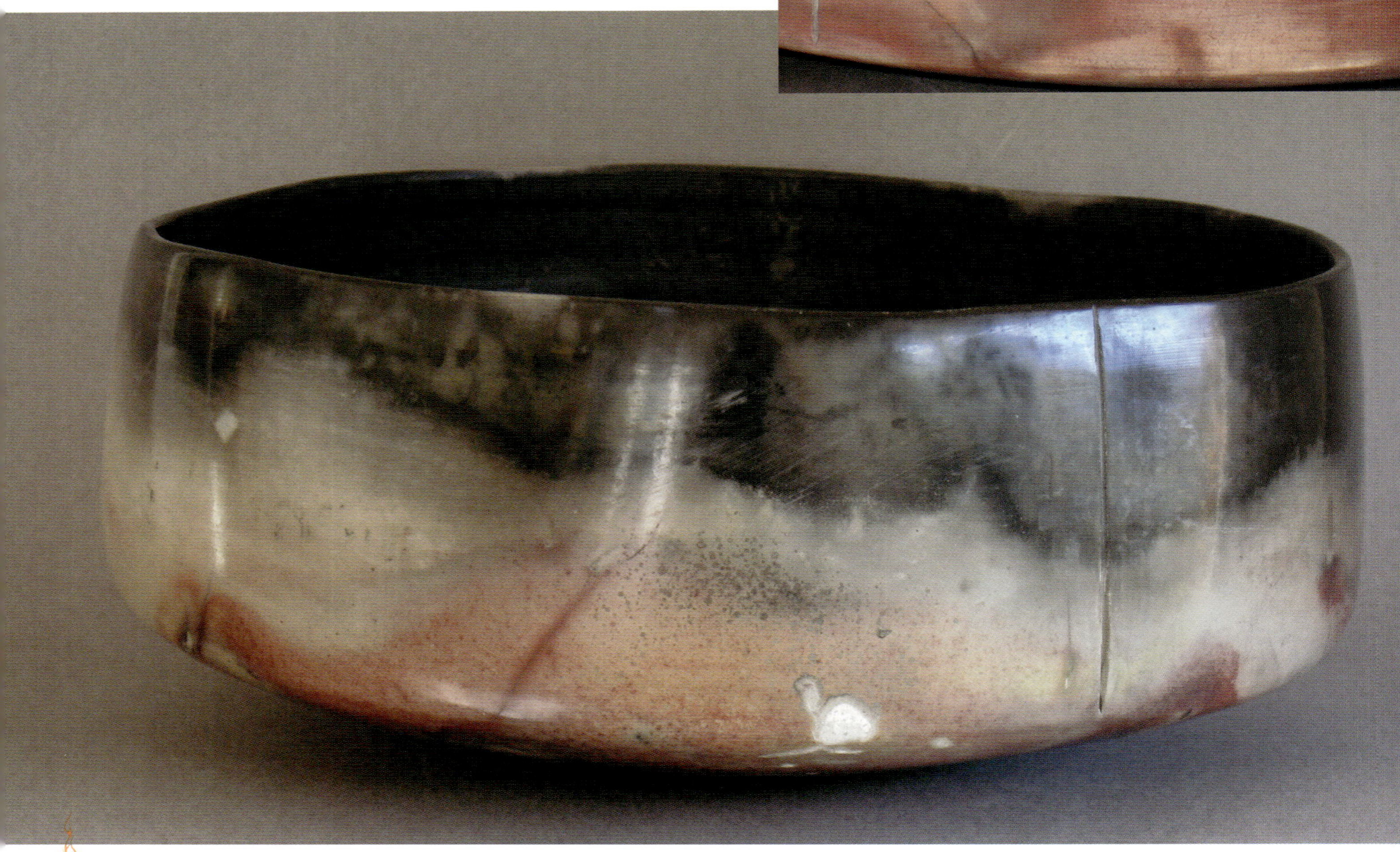

RAY ROGERS
(NZ RESIDING IN AUSTRALIA)

Ray Rogers' ceramics demonstrate a preference for simplicity in form, originating from a traditional background in ceramics, beginning forty years ago. Most of Rogers' work is based on the vessel, often with a contrasting worked area and mostly organic in origin, which adds interest and sometimes a challenge for the observer. Firing techniques employed since 1980 have relied on the vagaries of the fire itself, the movement of the flame manifesting as decorative surfaces of the forms—firstly using pit firing and more recently by post firing "Inglaze Silver Lusters" in a muffle kiln (Bow 1992, 49-51). Most of Rogers' works are wheel-thrown, and because many of his forms are quite large he mainly uses his own clay mixture, which he makes by adding grog and coarse fireclay to lessen the effects of thermal shock (Rogers 1983, 4). Rogers burnishes his ceramics, not to achieve the high gloss surface of the Native North American Indians and Mexican potters, but simply to attain a smooth surface that will compliment the subtle tonal hues that are accomplished through pit firing: he calls this semi-burnishing (Ibid).

Ray Rogers says of his artwork:

> The method of pit firing, being very immediate and risky in terms of survival rates, was useful in removing any notions of "preciousness" about my work. The simplicity of the forms accentuates the activity on the surfaces.
>
> Having been raised on the South Island of New Zealand in the village of Waitati on Blue Skin Bay by the sea, I have a strong interest in nature. What I have observed are the similarities between flora from the forest floor, for example fungi, and flora from the sea, for example coral. The fungoid carved design used on some of the work, manifested after a prolonged search for a method of resolving large spherical forms. There is a clear recollection that during "a moment of stillness" the resolution became obvious and thus demonstrates how everything is available to us "in the present moment," if we are open to it.

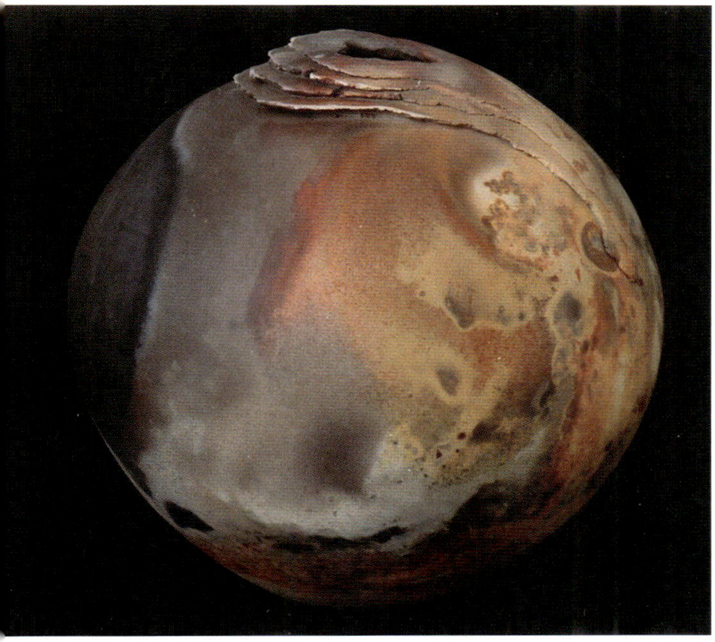

Ray Rogers. *Rock Formation*. Wheel-thrown, burnished, applied detail, pit fired ceramic. 12" h x 10 ½" w. *Photo: Ray Rogers*

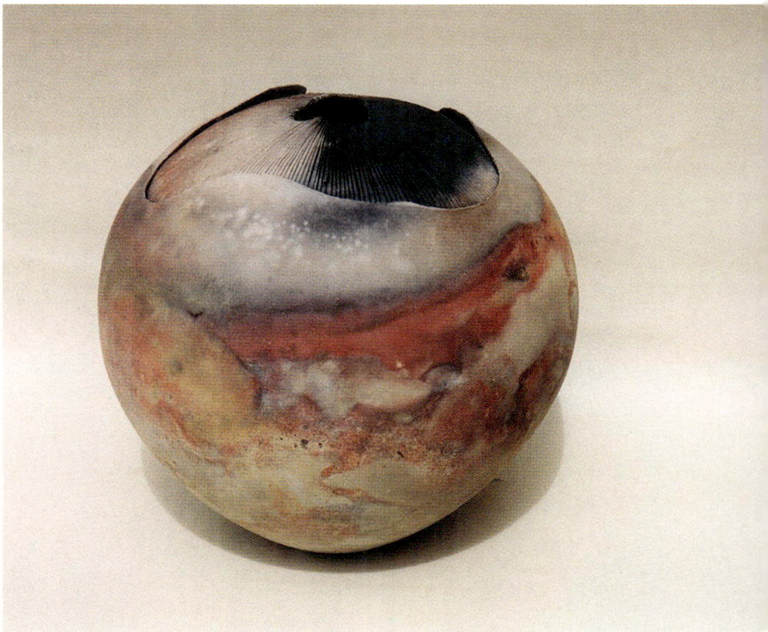

Ray Rogers. *Fungoid Form*. Wheel-thrown, burnished, incised, pit fired ceramic. 12" h x 10 ½" w. *Photo: Ray Rogers*

TANIA RUSTAGE

(US)

Tania Rustage creates her one-of-a-kind pottery pieces on the potter's wheel and then alters the form to make them unique, often adding wooden handles to heighten the true beauty of nature in its most natural form. She also enjoys making forms with an African Influence, and creates her artwork with the "individual" in mind.

Tania Rustage says of her art practice:

> These creations are all designed to be enjoyed as "Decorative Works of Art," known as "Smoke Fired Art," where the fire paints its design on the unglazed silky surface. My love of pottery comes from the use of natural earthy products, especially in alternative firing such as "Pit Fired" and "Saggar Fired" creations. Each piece has a look all its own, giving the piece depth and drama—the clay comes from the earth and the glazing is all natural, as it is created from the unpredictable beauty of the flames. No two pieces will ever be the same, and this is what draws me to this kind of work, as they are always unique. Each piece can be admired from every angle as the flamed design rotates around the form, and this is what draws the viewer into the artwork.

Tania Rustage. *Global Modification.* Wheel-thrown and burnished, terra sigillata. Bowl: 11.75" diameter x 3.25" h; Wood: 11.75" w x 6" h. *Photo: Tania Rustage*

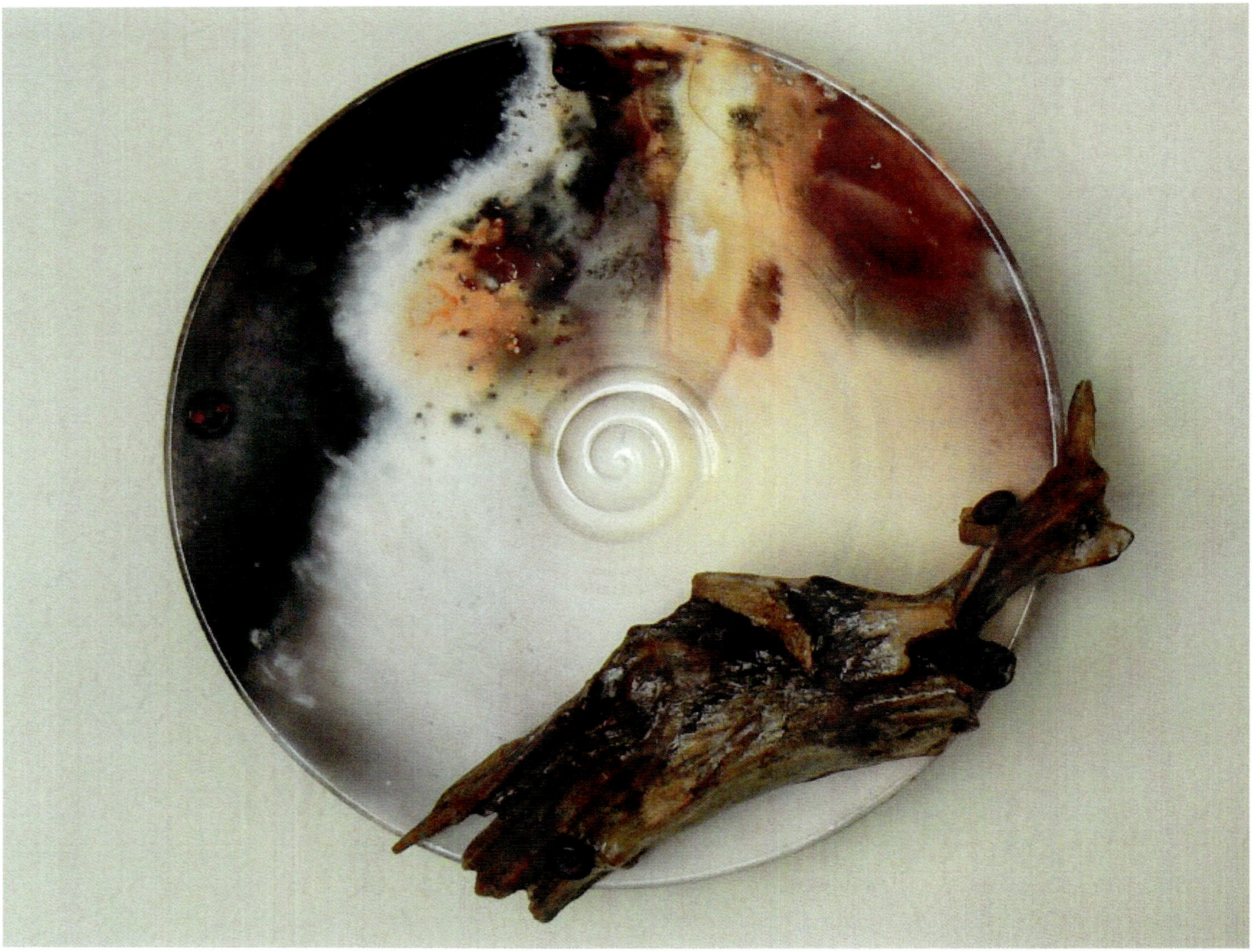

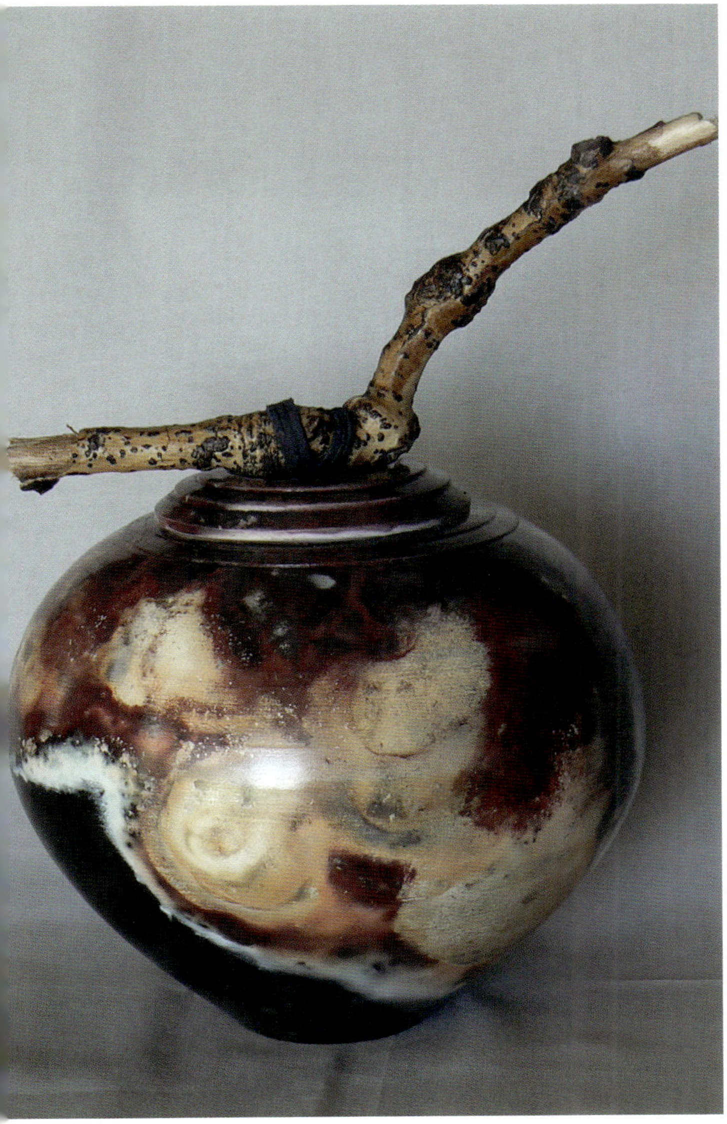

Tania Rustage. *Gentle Suppression*. Wheel-thrown and burnished, terra sigillata. Pot: 7" h x 7.75" d; Wood: 1.25" h x 8.25" w. *Photo: Tania Rustage*

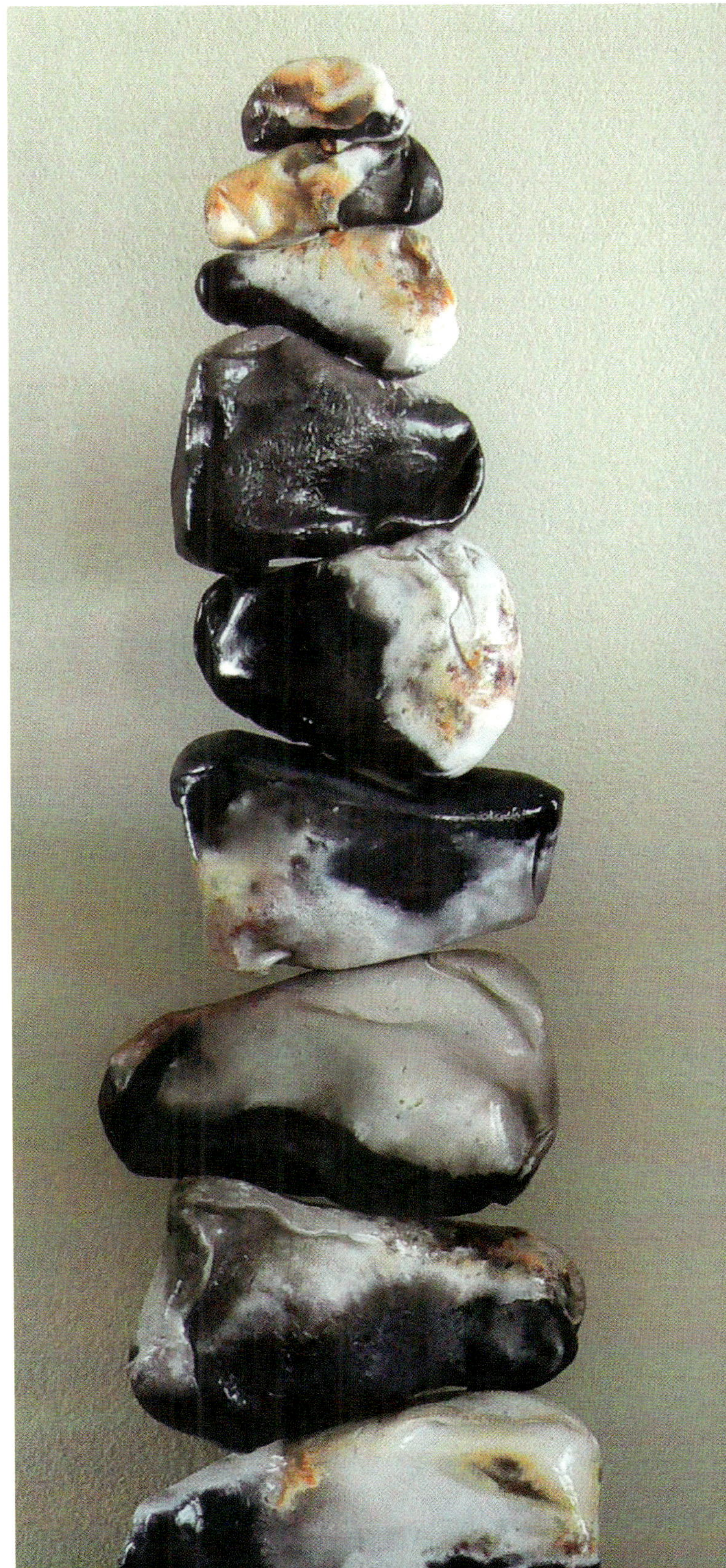

Tania Rustage. *Rock Totem* Handbuilt and burnished, terra sigillata. 50" h x 16" w x 9" d. *Photo: Tania Rustage*

SUZANNE SCHMALTZ
(US)

Art, and the exploration of it, have been part of my life since my childhood. I cannot remember a time that art was not part of my world. My mother, an artist, made certain that all forms of art and music were made available to her children throughout our lives.

My exploration of ceramics began in 1977 and I renewed my passion for ceramics in 2000 at college where I was introduced to all types of building and firing methods, including pit firing. Of all firing methods, pit firing is the only one that has endured and is the only finishing method that I utilize in my work.

Upon retiring in 2006, I was able to pursue art full time, beginning in 2008. The years 2008 and 2009 were spent building up my current body of work as well as establishing the business foundation on which to market and sell my work.

Why pit firing? For me it is the final expression between nature and humanity's desire to replicate it, with nature in command of that final expression. I love the fact that you can utilize the same materials as nature: temperature, humidity, and wind. Although the same clay body, chemical enhancements (oxides, sulfates, etc.), and firing materials are used, no piece ever emerges from the flame the same because no two days are ever the same. Your work is at the final mercy of nature's elements. For me, this represents a full circle interaction with the earth. We draw the clay from the earth, mold it to express ourselves, and give it back to the earth for her final thumbprint. A well fired piece feels like a blessing from the earth. Nature has approved of our expression of what we see and feel.

While I appreciate a beautiful ceramic piece that can begin and end in its singularity, for me, it is but part of a process. My ceramic pieces are my foundation on which to build a final vessel, which incorporates pine needle basketry as well as natural and found objects. Each piece is like the building of a nest and no two pieces will ever be the same, even though I incorporate many of the same materials in each of my pieces: like nature, with the same components, but expressed so differently in their final form.

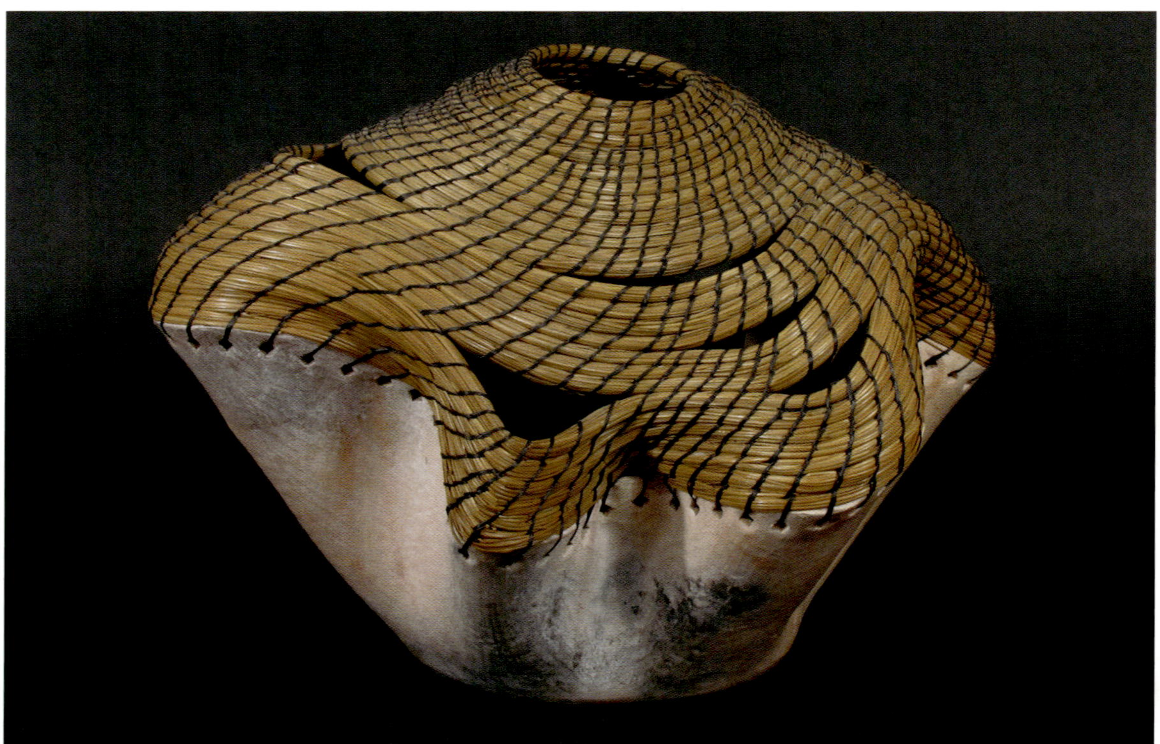

Suzanne Schmaltz. *The Start.* Handbuilt, burnished, pit fired ceramic, pine needles, artificial sinew enclosed basket. 19" w x 12" h. *Photo: Chris Everett*

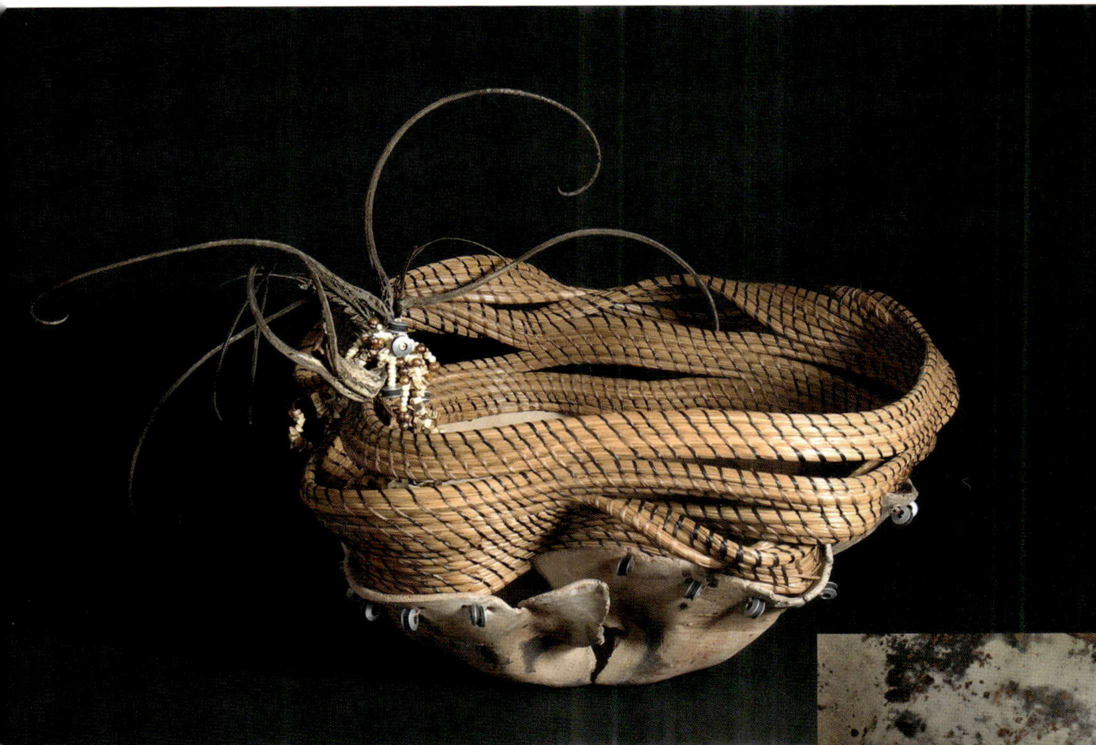

Suzanne Schmaltz. *While Searching for Zen.* Handbuilt, pit fired ceramics, pine needles, artificial sinew, glass and stone beads, found objects and devil's claw. 17" w x 14" h. *Photo: Suzanne Schmaltz*

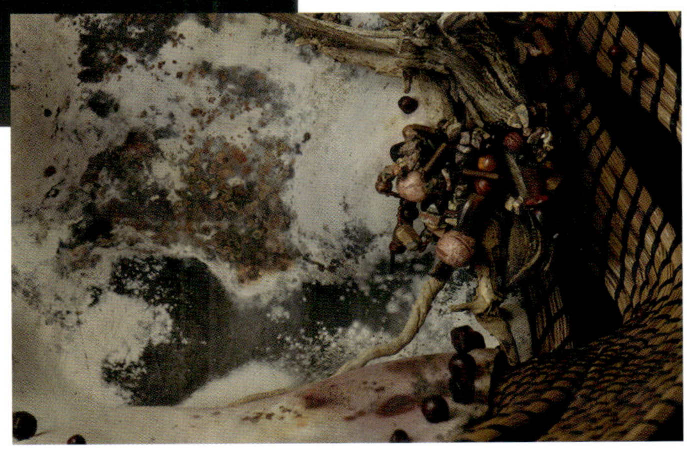

Suzanne Schmaltz. *While Searching for Zen.* Detail. *Photo: Suzanne Schmaltz*

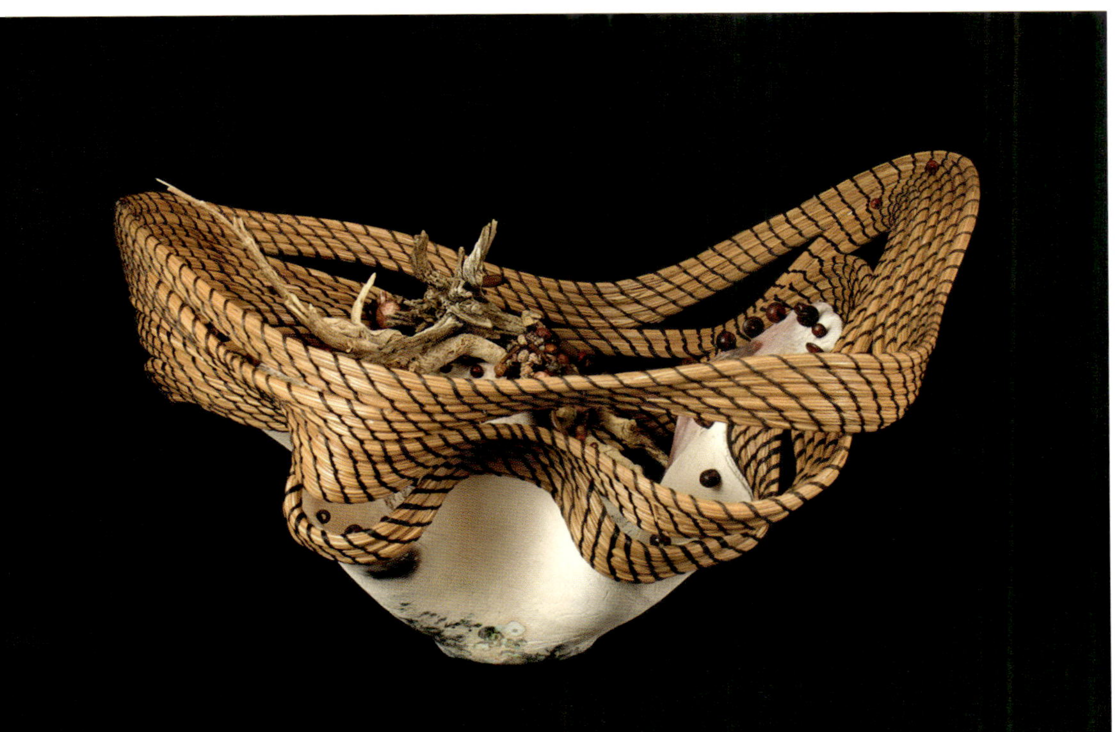

Suzanne Schmaltz. *After the Storm.* Handbuilt, pit fired ceramics, white clay body, heavy iron oxide influence, pine needles, artificial sinew, clay, glass, stone, and other beads and found objects. 21" w x 10 ½" h. *Photo: Chris Everett*

UP IN SMOKE POTTERY
(US)

Chad Dykstra is a self-taught potter specializing in pit and saggar firings, creating unique decorative pottery. Growing up in rural South Dakota, life moved at a slower pace. Dykstra spent hours looking at the sky, whether it was the stars at night or a thunderstorm roll in. He was always mystified at the perfect randomness he found in nature.

All of his pieces are finished without glazes in either a pit or saggar firing process. These traditional firing methods give the finished pieces a natural appearance and visually stimulate sights from his childhood, providing a strong connection to the artwork.

Together with his wife, Kiesha, they create and sell their pottery in the Upper Midwest at select galleries and summer art festivals.

Of his inspirations and art practice Dykstra says:

> Pottery has become the medium to let the natural instinct of my artistic self be expressed. I see each vessel conveying its own entity, which through molding becomes the creation it is meant to be. I rarely have a complete vision for each piece as I sit at the wheel to create, they are formed as a collaborate effort between myself and the clay. The form captures my mood, my thoughts, and my emotions at the time it is made. I purposefully smooth and polish my vessels prior to firing; this surface preparation assists in the color and pattern formation in the firing process.
>
> I became interested in primitive firings, first from an economic point. It evolved into a desire to see how the placement of different metals, salts, minerals, or organic materials around the piece would affect the pots during the firing. Each one of these "elements" reacts with the clay differently with each firing. Each firing produces unique results; thus each piece is individual.
>
> After every firing, the reward of examining each artistic piece allows me to see the colors and patterns that the firing produced; this demonstrates the creativity of the firing process. The finished pieces exhibit subtle layers of color, sometimes like ominous clouds. Others are sprinkled with vibrant color similar to photos of stars and far off celestial bodies. It wasn't until several years ago that I realized the important role of these images in portraying to me the vast nighttime sky above the prairie or the ominous thunderclouds. These images were the visions of home, where I grew up and the sights I saw on a regular basis. This is the reason I continue to use these methods to establish the voice of my artwork.

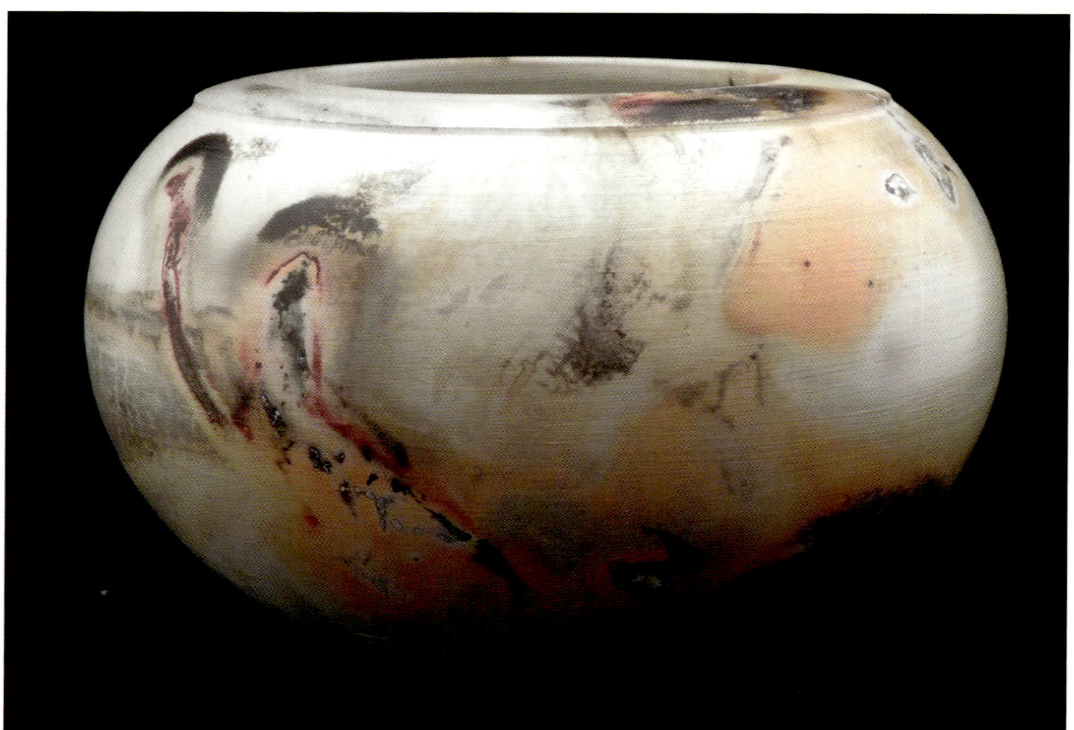

Up in Smoke Pottery. *#432.* Wheel-thrown stoneware clay body, pit fired. 5" h x 7.5" w. *Photo: Chad Dykstra*

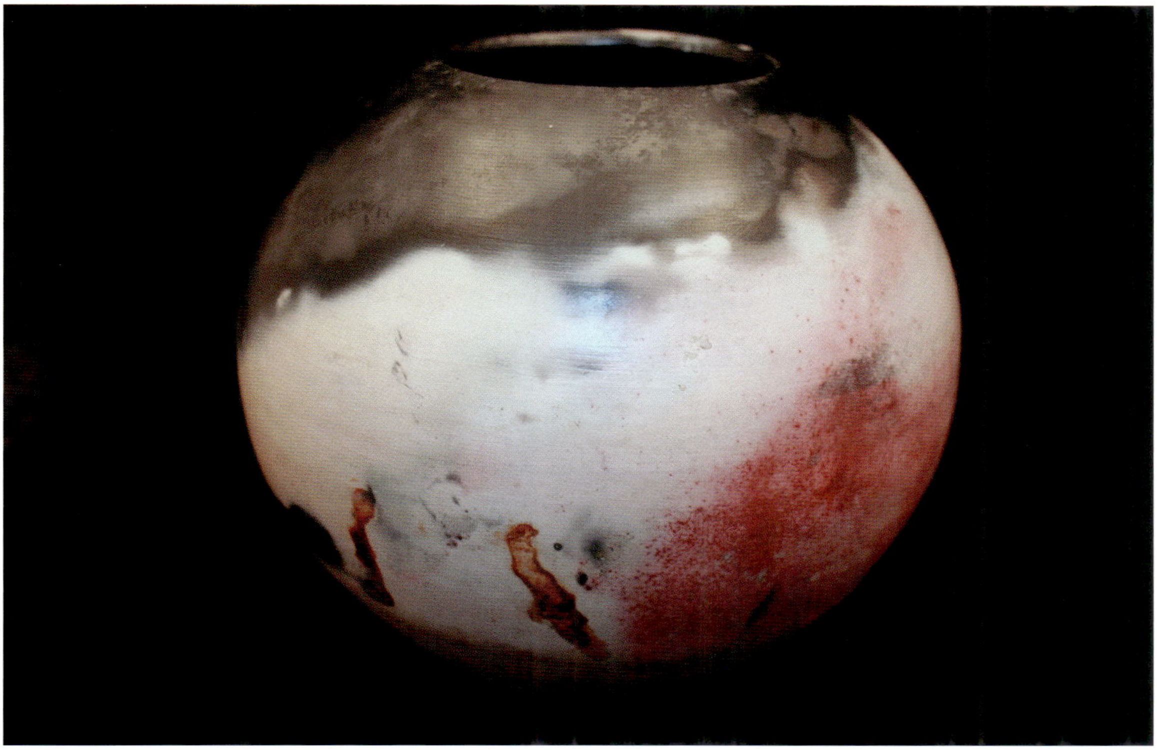

Up in Smoke Pottery. *#177.* Wheel-thrown raku clay body, pit fired. 9.25" h x 9.25" w. *Photo: Chad Dykstra*

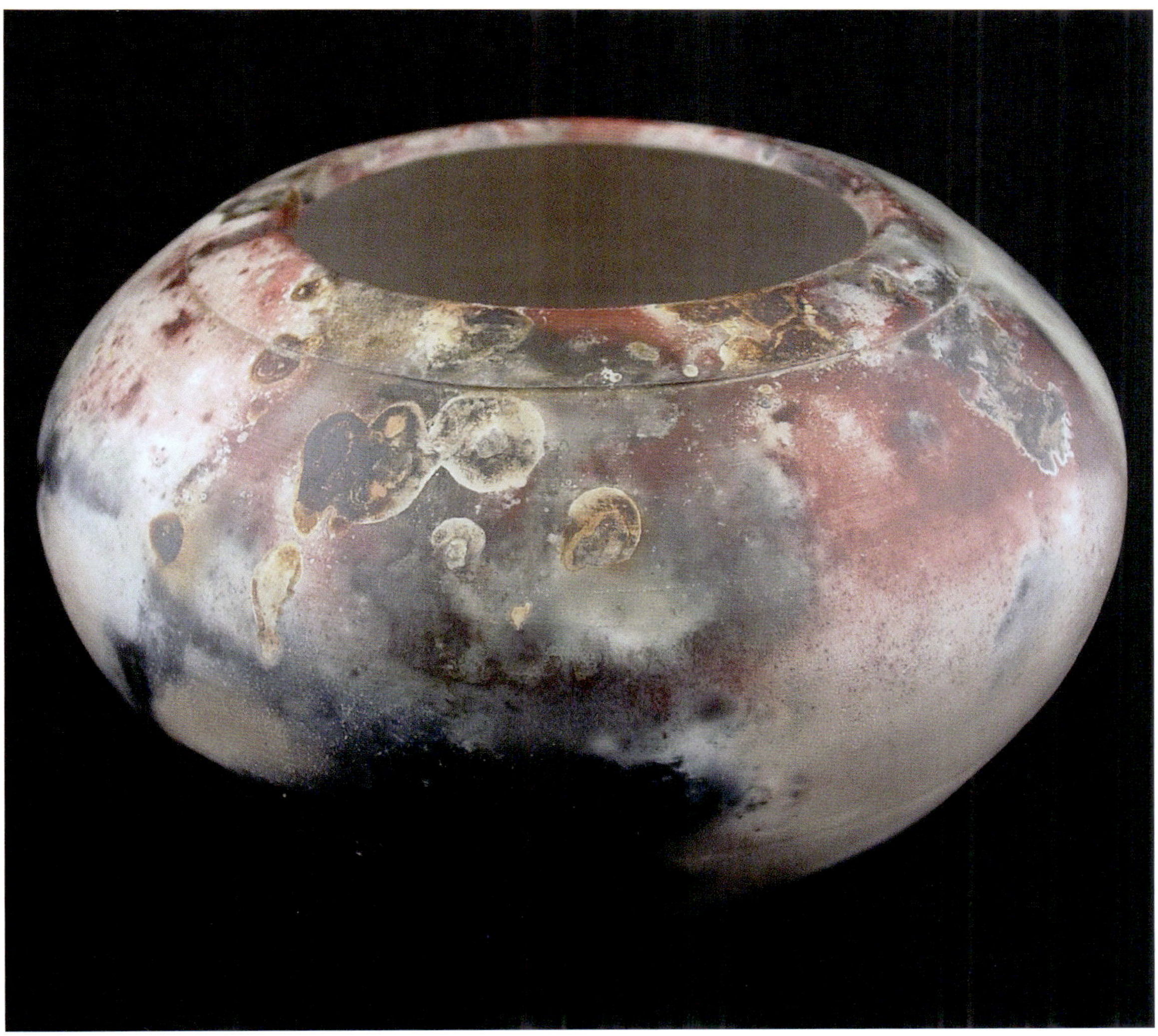

Up in Smoke Pottery. *#222.* Wheel-thrown stoneware clay body, pit fired. 6.75" h x 9" w. *Photo: Chad Dykstra*

SUMI VON DASSOW
(US)

Sumi von Dassow was born in Seattle and began studying ceramics in high school. She continued her ceramics education at the University of Washington and later at San Francisco State University, where she graduated in 1987 with a degree in Art with an emphasis in Ceramics.

Von Dassow now lives in Colorado, and is a well known potter and ceramic artist, teacher and writer. Her work includes brightly colored functional pottery, burnished and pit fired vessels, and ceramic sculpture. Von Dassow's ceramic work has won numerous awards in major local and national shows. Her art has appeared in books and magazines, and is in the permanent collections of several arts institutions in the country.

In addition to creating her own innovative work, von Dassow teaches pottery at Lakewood's Washington Heights Art Center, and has led numerous workshops around Colorado. She frequently contributes articles to the magazines *Ceramics Monthly* and *Pottery Making Illustrated*, and writes a book review column for the latter magazine. She edited the books *Barrel, Pit and Saggar Firing* in March 2001, and *Electric Kiln Pottery* in February 2003, both published by the American Ceramic Society. Her book *Low-firing and Burnishing* was released in 2009, as well as a DVD entitled *Pit Firing and Burnishing*.

Of her work Sumi von Dassow says:

> I was first exposed to pit firing in college, firing on the San Francisco beach with David Kuraoka. I fell in love with burnishing, fascinated with being able to make such a rough and earthy material take on a mirror-like surface. I have always felt that the burnished surface is warm and sensual, compared to the cold glassy surface of a glazed pot. I love working with clay, collaborating with it to achieve whatever it is capable of—and while I love using glazes too, I do sometimes feel that a coat of glaze is a barrier between me and the clay.

Pit firing is a collaboration between the fire, the air, and the materials I put in the pit. A potter who pit fires is working with all the traditional elements—earth and water to form the pot, fire and air to finish it. When I make a pot for pit firing, I have to give up some of the control I exercise in making it—and I have to make forms that will be enhanced by the results of the firing. I enjoy giving up a bit of control and working with the elements to create work that I could never duplicate by any other means. What other medium offers such an opportunity?

After graduating from college, I began teaching at the Washington Heights Art Center in Lakewood, a city-owned community art center. I had no particular thought of pit firing. I continued making burnished work and smoking it in my electric kiln to blacken it. My experience with pit firing was in a large pit, on the beach, with dozens of participants, and I didn't expect to find that experience in Colorado. However, my students started bugging me to find a way to do pit firing and we did happen to have a large gravel covered firing yard that was perfectly suitable—so with their help I embarked on a several year long process of discovering how to make pit firing work on a relatively small scale at high altitude.

I have been lucky in several respects—instead of burnishing all our pots with stones, as I do for my black pots, I decided to explore terra sigillata to make it easier for my students—and it turns out that terra sigillata made from OM-4 ball clay is crucial to developing good color from salt in my pit. The clay I usually choose to work with, B-mix from Laguna, does not get good color when it is burnished with a stone. And for fuel I rely on aspen wood, which burns hot and fast and leaves little ash or pitch on the pots, and which is abundant and easy to find and collect in Colorado. While many potters use seaweed in pit firing as a way to introduce salt into the firing, I experimented with soaking various materials in salt water. I now always use corn husks and pine needles, two materials that are easy to obtain for free, and keep forever—although I now usually have seaweed for firing also, since whenever I or any of my students travel to the coast we always bring back as much seaweed as we can!

So, through life's twists and turns, I have ended up with a passion for the vagaries of this firing method that I can't quite control, but that produces such spectacular results when all goes well. When I get frustrated with glazes that run and pinhole, with pots that are supposed to hold water and unexpectedly leak, with kilns that shut off prematurely, and with bargain-hunting customers, I find relief in pit firing! On the other hand, when I get frustrated at the fragility of pots that are only fired to cone 010, with terra sigillata that peels off, at ash and salt forming rough patches on my pots, and at the challenge of selling non-functional pots, I find it a relief to go back to glazed functional ware for a while.

Well, the truth is I love working with clay and I love fire, so what could be more fun than pit firing?

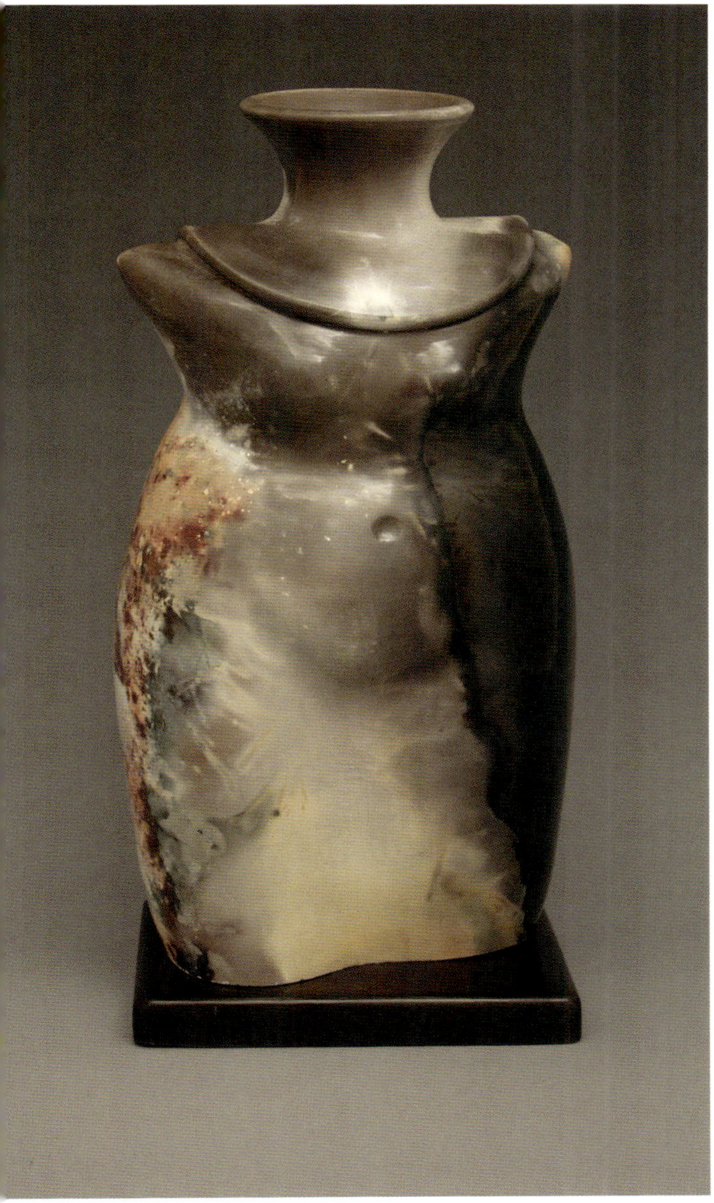

Sumi von Dassow. *Aphrodite.* Cone 5 B-mix, thrown and altered, with terra sigillata, bisque-fired to cone 010, then pit fired. 16" x 7 ½" x 5 ½". *Photo: Sumi von Dassow*

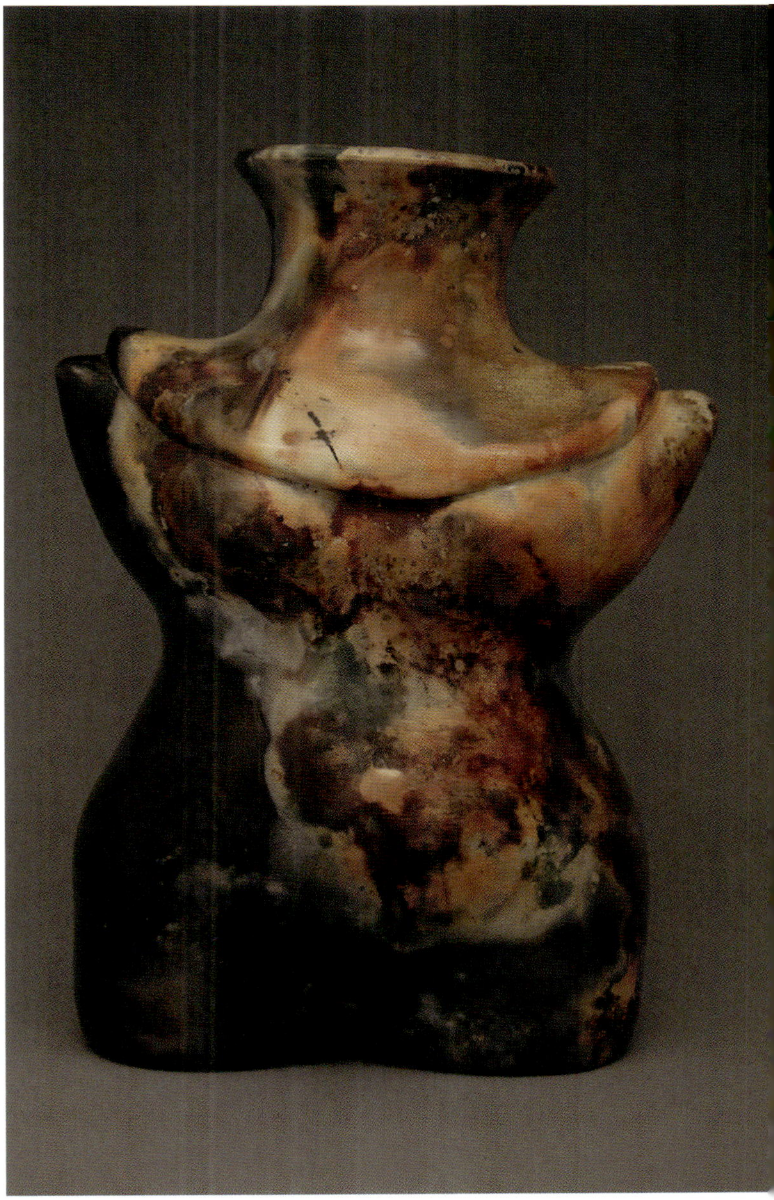

Sumi von Dassow. *Torso One.* Cone 5 B-Mix, thrown and altered, with terra sigillata, bisque-fired to cone 010, and pit fired. 11" x 7" x 5". *Photo: Sumi von Dassow*

JAN WALLACE
(AUSTRALIA)

Jan Wallace creates her pots using the coiling method for large pieces and wheel throwing for smaller pots. Sometimes she combines both methods, depending on the form she is making. She uses white raku clay with .5 mm grog for the larger coiled pieces, for its ability to withstand rapid temperature in the pit, and has had success using a white earthenware clay for the smaller wheel-thrown pots. When her pots are just past the leather-hard stage, Wallace paints them with terra sigillata, then buffs them with a car polishing cloth to produce a smooth, shiny, burnished finish. Wallace uses a white terra sigillata, using the same clay body as the pot, by saving her scraps and turnings. The pots are then bisque fired to approx 1740–1790° F [950–980° C]. Recently Wallace has also been making a red terra sigillata made from locally found clay.

Of her ceramic practice Jan Wallace says:

I have always loved to be creative and ceramics gives me the balance I need in my life. I particularly love the endless possibilities of creating objects in clay. There are so many different things to explore with respect to clay, glaze, and the firing process itself, that my mind is constantly challenged. Each time I explore a new idea, it leads me to another exciting pathway and it is that process, the challenge, and the excitement that I love.

My love affair with clay started in high school over twenty years ago. I later pursued the craft after leaving school and set up a small studio in a pre-loved horse stable in our backyard at home. I bought myself a pottery wheel and attended a

Jan Wallace. *Bushfire Banksia.* Handbuilt, white raku clay, sprayed with locally found red clay made into terra sigillata, bisque fired to cone 06, iron ox wash, pit fired with salt and seaweed. 12.5" w x 7.8" h. *Photo: Jan Wallace*

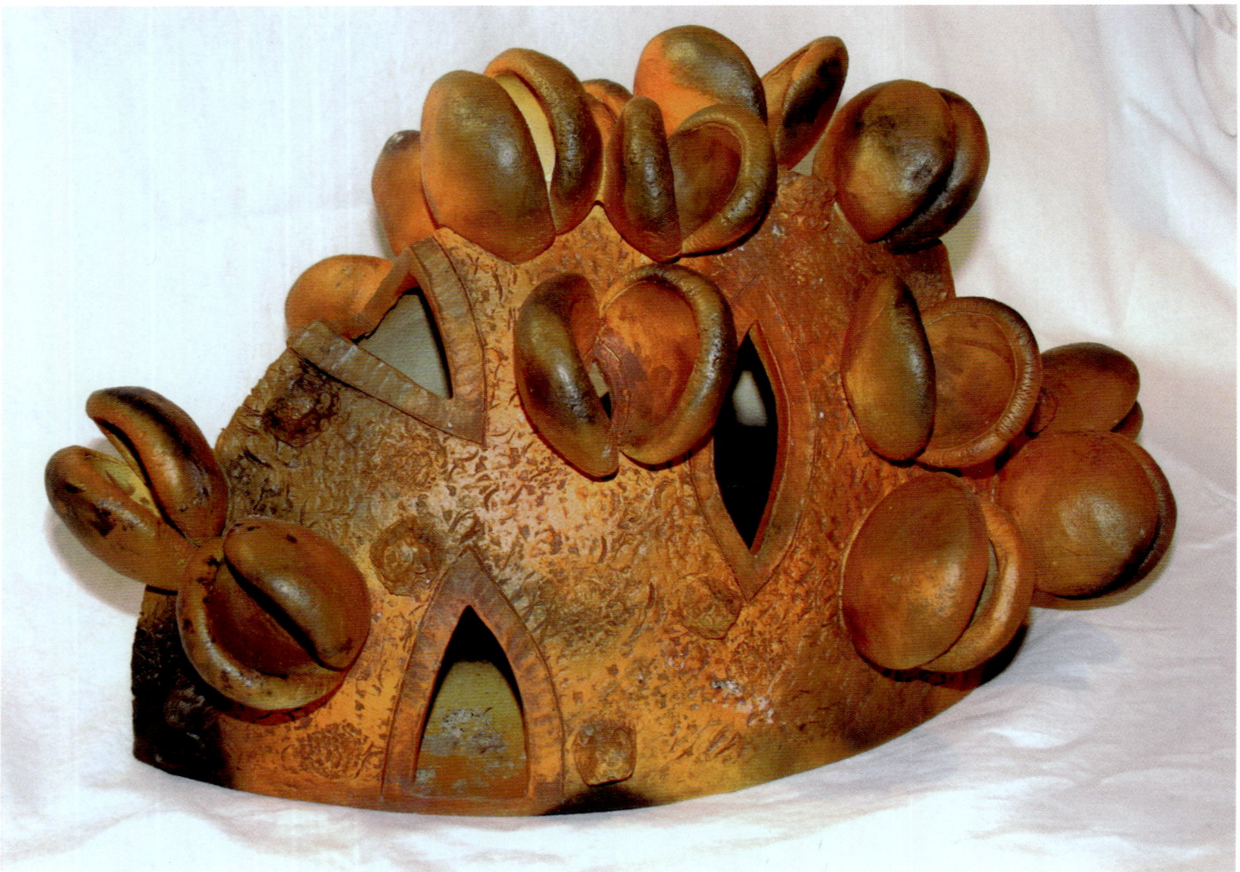

local, weekly pottery class for a year. I then decided to enroll in a TAFE course, and over the next three years I attended Hornsby TAFE in NSW and completed Ceramics Certificate III and Surface Treatment & Design (1999–2001). There was a short period in my life while living in Queensland, when I was unable to commit to ceramics due to my circumstances. Now, however, I am living back in NSW, and am enjoying the opportunity to have time to commit to my ceramics again. Recently I began to enter competitions and approach galleries and as a result was recently invited to feature a series of my raku work called Sea Shells in the 32nd Anniversary Exhibition at Articles Fine Art Gallery in Stanwell Park, NSW.

I am a versatile person and enjoy making a wide range of ceramics. I am particularly interested in making one-off artistic pieces. Nature is a big source of inspiration for me as well as human made objects from the past (rusty objects like corrugated iron, cogs, wheels, etc.). I love the process of raku and pit firing. The simple, beautiful, and somewhat unpredictable results that are achieved by firing in these ways keep me coming back for more. The whole process of pit firing is exciting and almost ritualistic. When pit firing, I am able to continue the connection with my work in the firing process until the final moment when a piece becomes cool enough to hold.

As in all areas of ceramics, there are highs and lows with pit firing. There are pots that make your heart sing when they come out of the pit and others that are a complete failure. I personally don't mind the odd failure. I have enjoyed the process itself, I have learned something valuable during the process, and once a pot is completely finished I am looking forward to the next part of the journey.

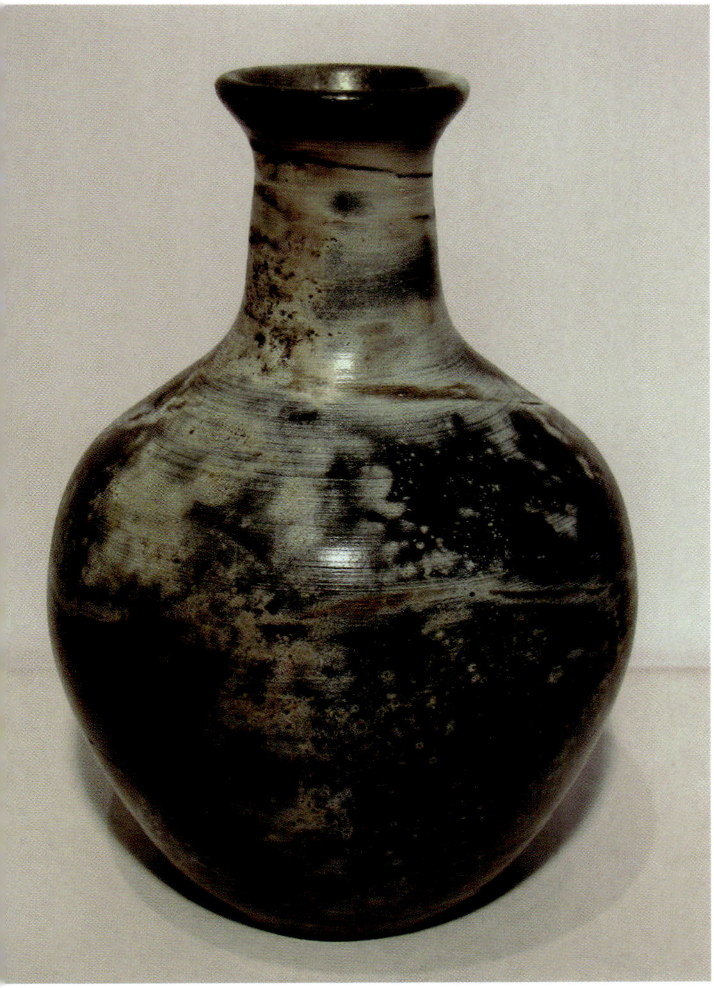

Jan Wallace. *Pit Fired Bottle Form*. Wheel-thrown white earthenware, painted with white earthenware terra sigillata, bisque fired to cone 06, copper wire tied around neck of piece and wrapped in an aluminium saggar with paper, sawdust, copper carbonate, seaweed, and salt. 6.2" h x 3.9" w. *Photo: Jan Wallace*

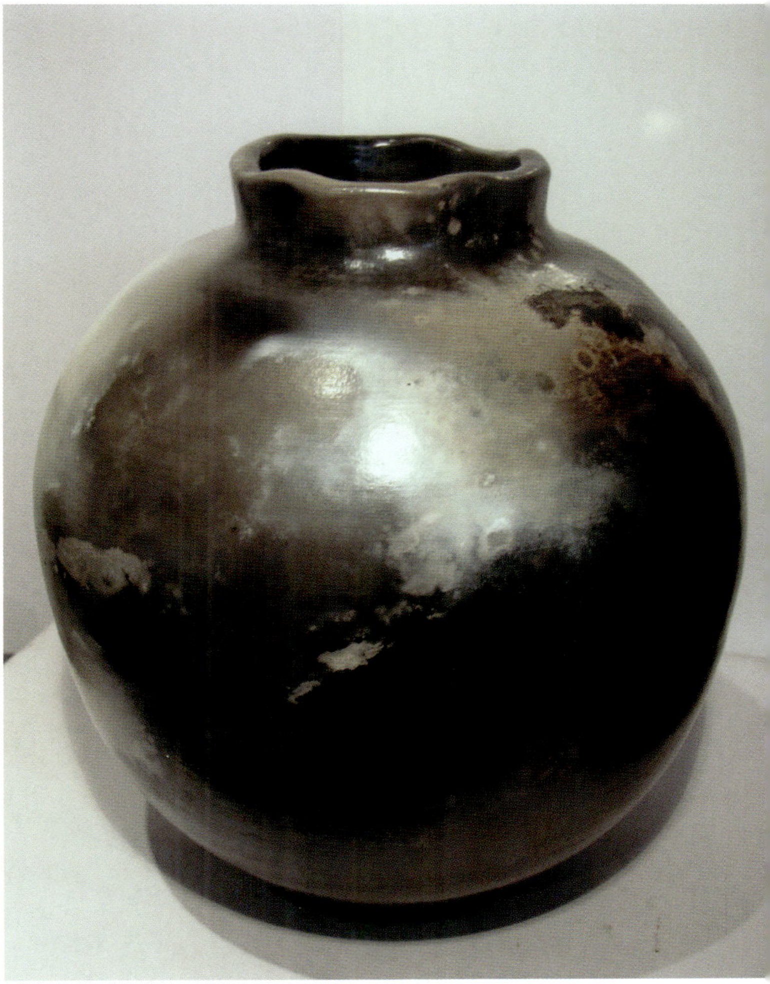

Jan Wallace. *Pomegranate Seed.* Wheel-thrown white earthenware, painted with white earthenware terra sigillata, bisque fired to cone 06, wrapped in an aluminium saggar with paper, sawdust, copper carbonate, seaweed, and salt. 11.8" h x 11" w. *Photo: Jan Wallace*

MICHAEL WEIN
(AUSTRALIA)

Michael Wein has been a self taught potter for twenty-eight years, specializing in Japanese style raku domestic ware and pit fired artistic pieces. His inspiration comes from studying Japanese pottery and from personal visits to, and the lectures of, Japanese Master Potters. Wein has exhibited widely in Australia, Japan, and the USA, has undertaken several residencies in Japan, and most recently undertook a residency in Aspen, Colorado. His work is held in gallery collections in Australia, Japan, and the USA. Wein's pit fired ceramics are fired in a 13' l x 6' w x 4 ½' d pit dug into the ground. He uses a variety of ingredients to achieve his results, including copper, cobalt, and Neptune bead seaweed. Some of Wein's works are then hand carved to capture the Pectinia coral that is found along the coast where Wein lives. Carving these pots takes a full day.

Images of Wein's work were recently published in *Ceramics Today* (Schiffer Books) in 2010.

Of his work Michael Wein says:

> The Spirit of Pit Firing:
>
> Be free. There is a necessity to embrace the element of surprise that a piece painted by fire can give. I do not fear to lose some pieces and what was once planned. I have an urge to grow with my discoveries. Embrace the unknown. I make no demands and accept what the fire and nature gives me. I follow no absolute plan and will accept another solution. I am happy to gamble with some pieces and for my intuition to grow.

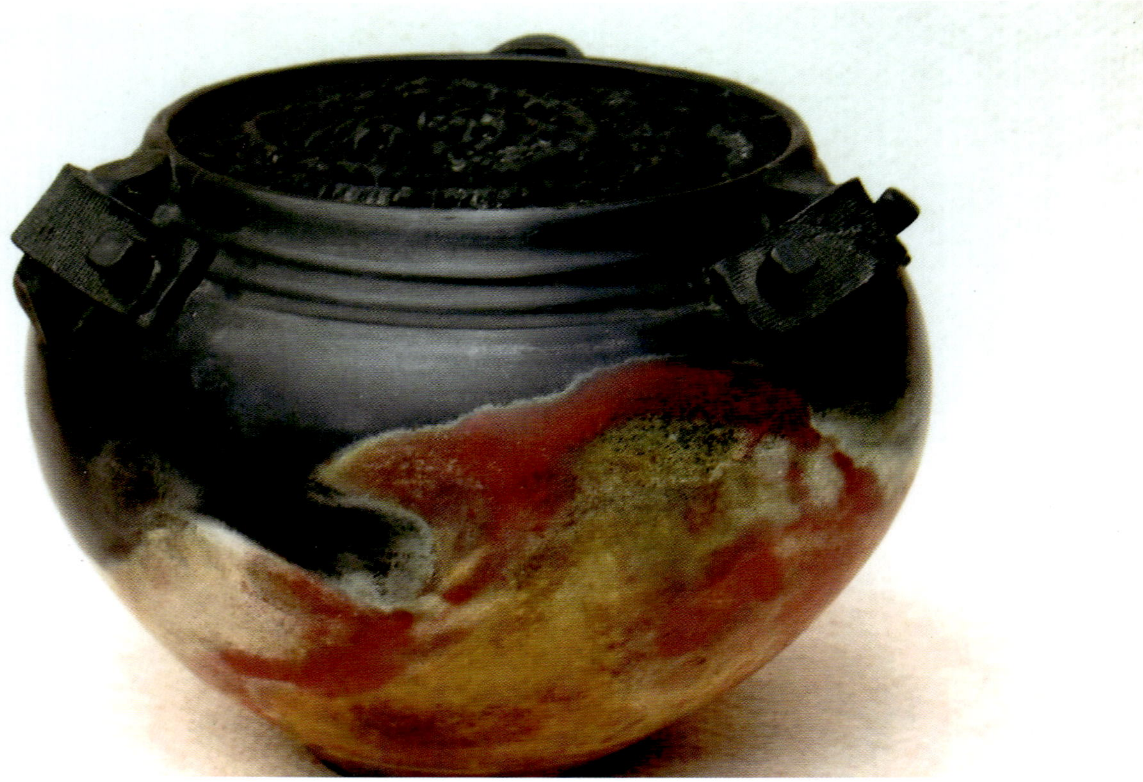

Michael Wein. *Untitled 1.* Wheel-thrown, carved top, burnished, pit fired upside down. 15" h x 14 ½" w. *Photo: Michael Wein*

Michael Wein. *Untitled 2.* Wheel-thrown, burnished, pit fired with copper and Neptune bead seaweed. 19 ¾" h with stand x 12 ½" w. *Photo: Michael Wein*

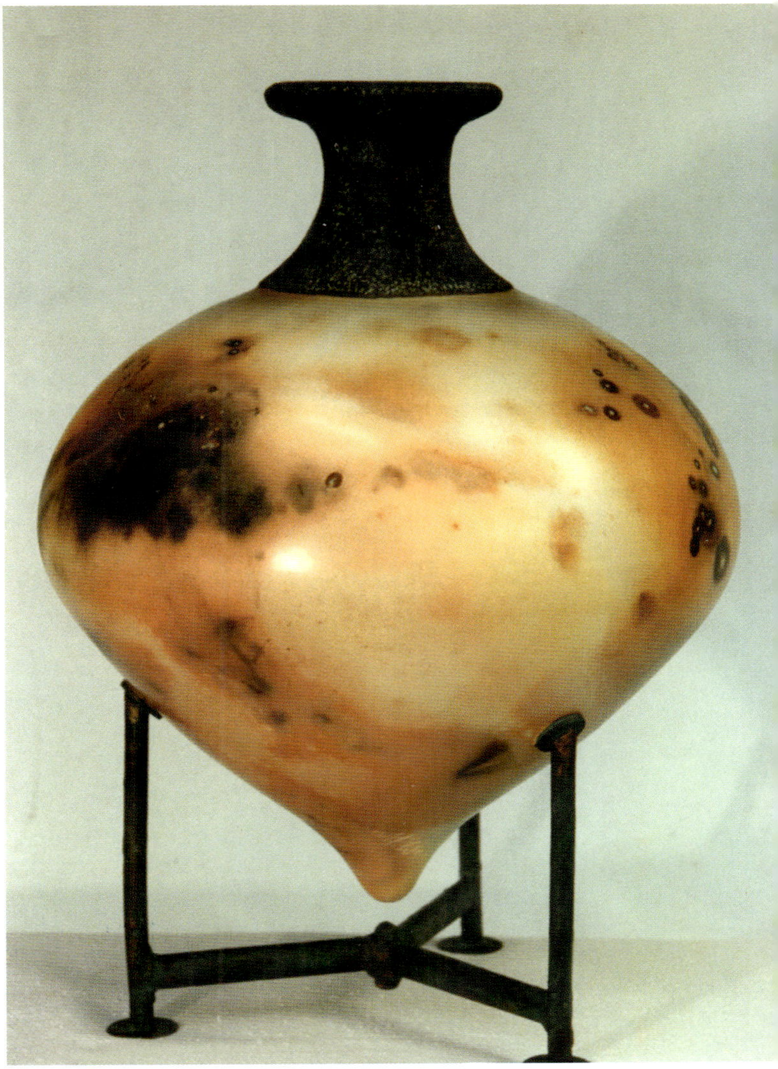

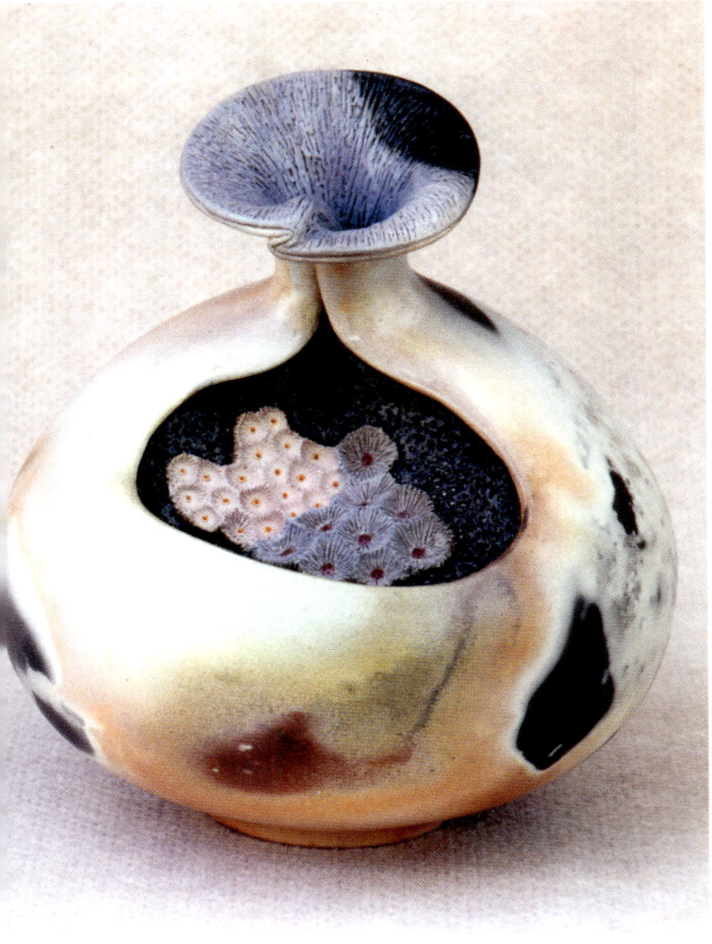

Michael Wein. *Untitled 3.* Wheel-thrown, burnished, carved insert. 14 ¼" h x 11" w. *Photo: Michael Wein*

JIM WHALEN
(US)

Jim Whalen's pots are wheel-thrown, burnished, and coated with terra sigillata. After bisque firing, he creates patterns and images using a wax resist method. The pots then go through a two stage firing process. The first is a low-temperature salt firing, the second a pit firing using sawdust. Both of these firings produce random chaotic markings on the surface of each pot. The wax resist separates the effects achieved by each firing into distinct areas, giving definition to the patterns without losing the vitality of chaos.

Jim Whalen says of his art practice:

These pots represent my desire to unite the forces of nature with the forms of the subconscious, exploring and expressing the connection between the two. Salt, sawdust, and the iron in the clay produce the variety of earth tones on the smooth surface of the pots. The surface of the pot is reminiscent of polished wood or stone. The simple, rounded form combined with the dynamic surface elements results in a work of art that is universal and timeless.

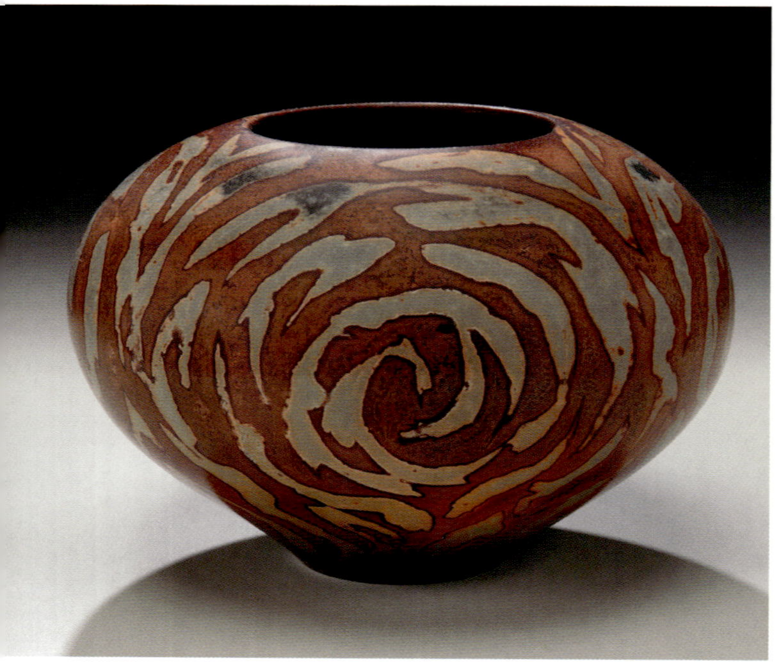

Jim Whalen. *Spiral Abstract.* Wheel-thrown, burnished, terra sigillata. 6" h x 9" w. *Photo: Tim Barnnell*

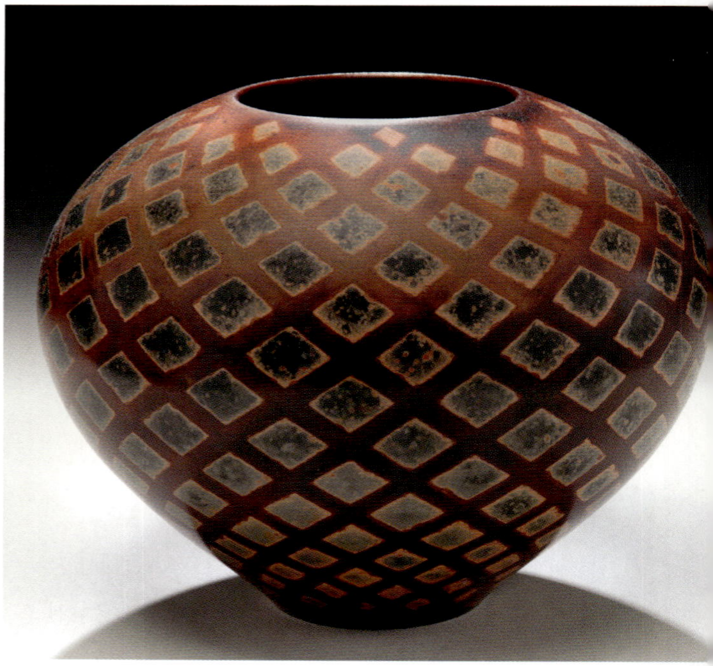

Jim Whalen. *Diamond Pattern.* Wheel-thrown, burnished, terra sigillata. 7" h x 9" w. *Photo: Tim Barnnell*

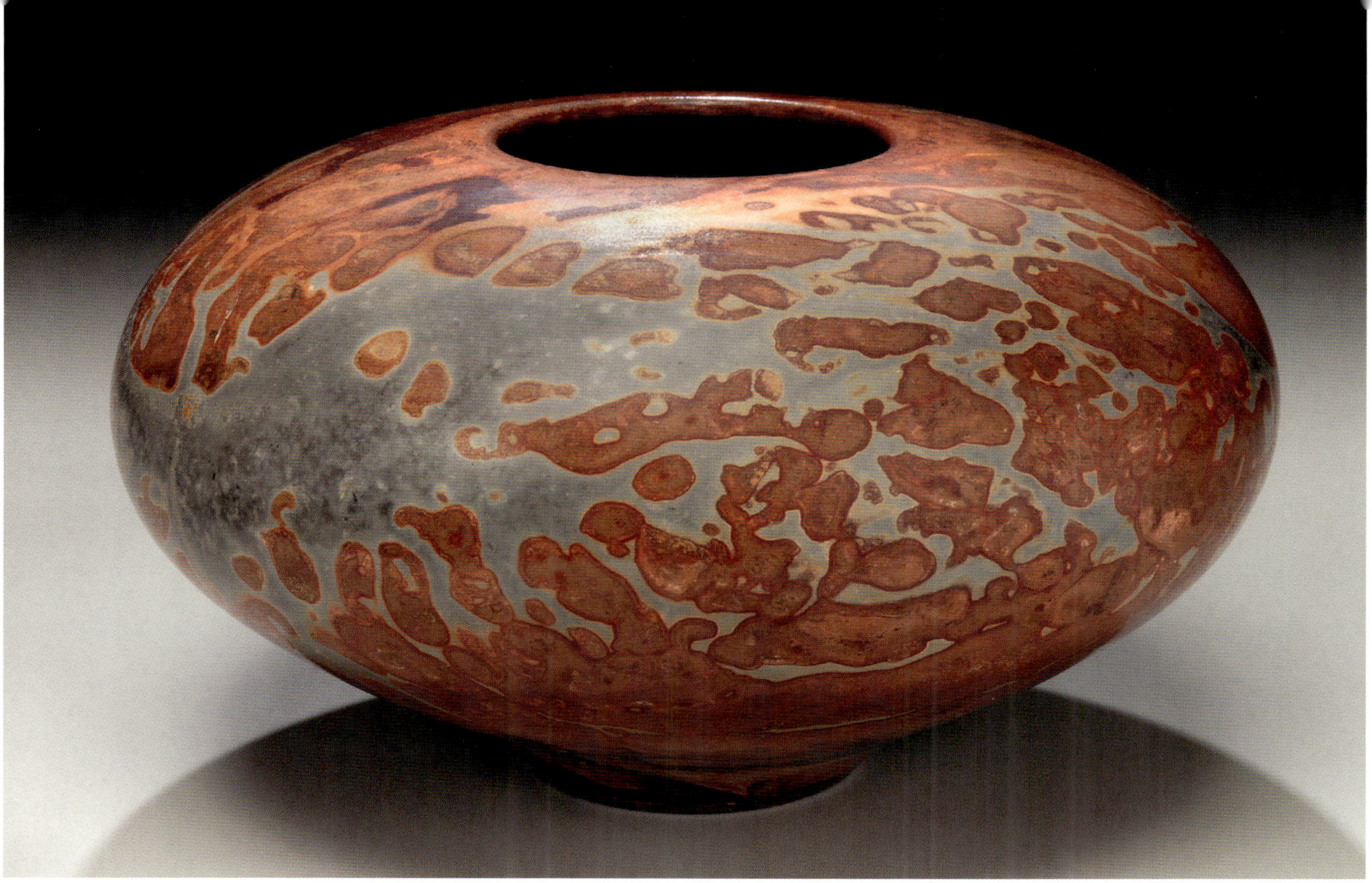

Jim Whalen. *Marble Pattern.* Wheel-thrown, burnished, terra sigillata. 8" h x 12" w. *Photo: Tim Barnell*

Jim Whalen. *Polar Pattern.* Wheel-thrown, burnished, terra sigillata. 6" h x 9" w. *Photo: Tim Barnell*

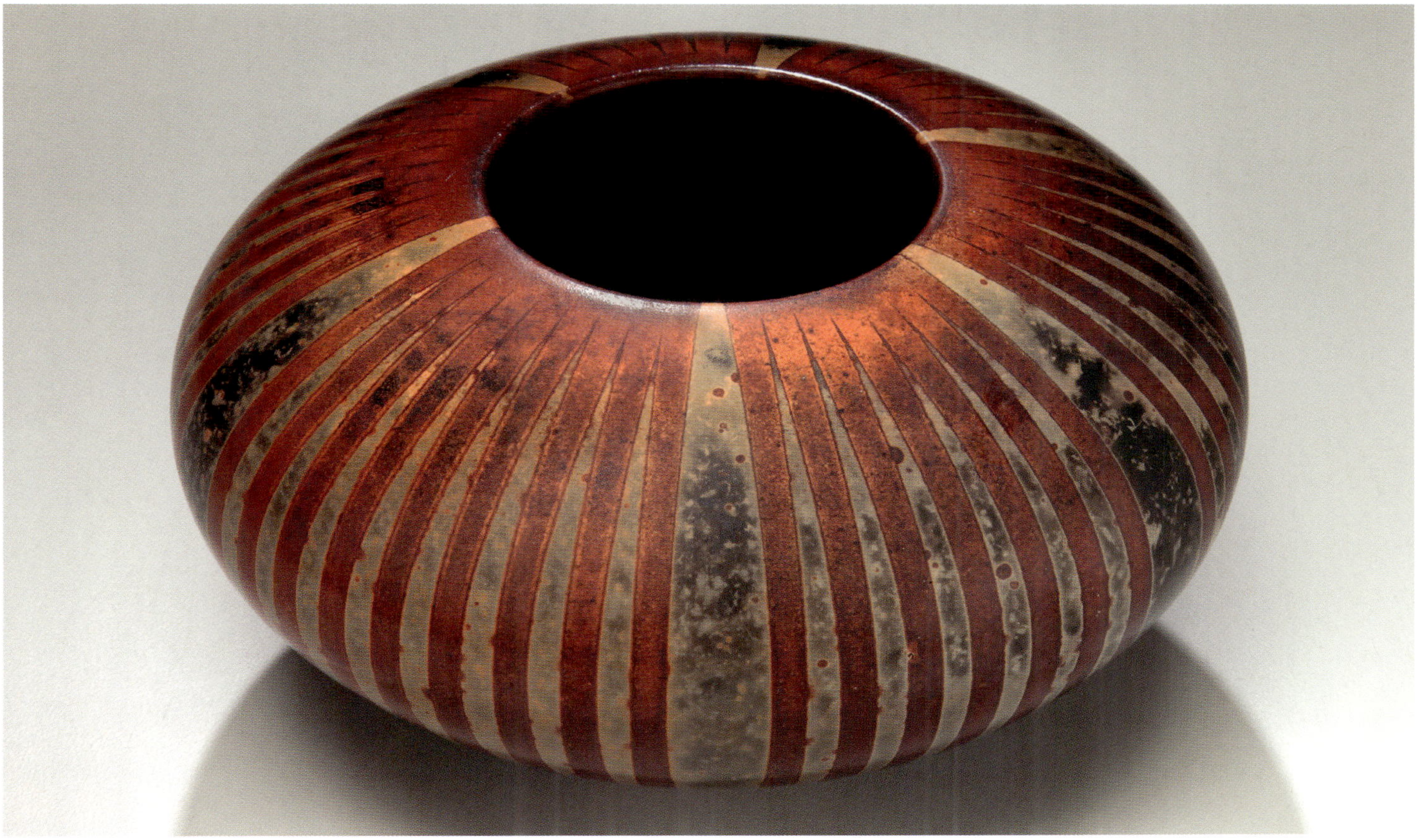

DAWN WHITEHAND
(AUSTRALIA)

I was originally introduced to clay as an adult via a hobby class at a local community center and felt an immediate affinity with the medium, and have now been a ceramic artist for many years. During this time I have exhibited widely both in Australia and overseas, undertaken artist residencies, had images published in books, published journal articles, and won several awards.

I discovered pit firing during my honors years at university. I spent my undergraduate years making functional ceramics, but I wanted to explore the potential of the vessel to communicate from a more conceptual basis. As I began to gradually move away from shiny functional glazes, I became attracted to the notion of firing in the ground. I live on a large property in regional Victoria, Australia, so there are no neighbors to be disturbed by the smoke or fumes.

I wanted to continue my exploration of the ceramic vessel, and went on to complete my PhD, and pit firing formed a large part of my methodology. My work became more organic in form and surface, and my hypothesis was that such artwork, when installed in the natural environment, can trigger innate responses in the viewer, enabling a reconnection to the earth. Such a reawakening, I believe, is crucial if humanity is to cease using the earth as a commodity.

I also combine my ceramic forms with natural and manufactured objects that are organic in shape or texture, creating mixed media artworks. These sculptures are suitable for the gallery context, yet I still create them with the same theme in mind: to regenerate an earthly connection within the viewer.

Images of my work have been published in *500 Ceramic Sculptures* (Lark Books), *500 Raku* (Lark Books), *Ceramics Today* (Schiffer Books) and most recently, *The Best of 500 Ceramics* (Lark Books).

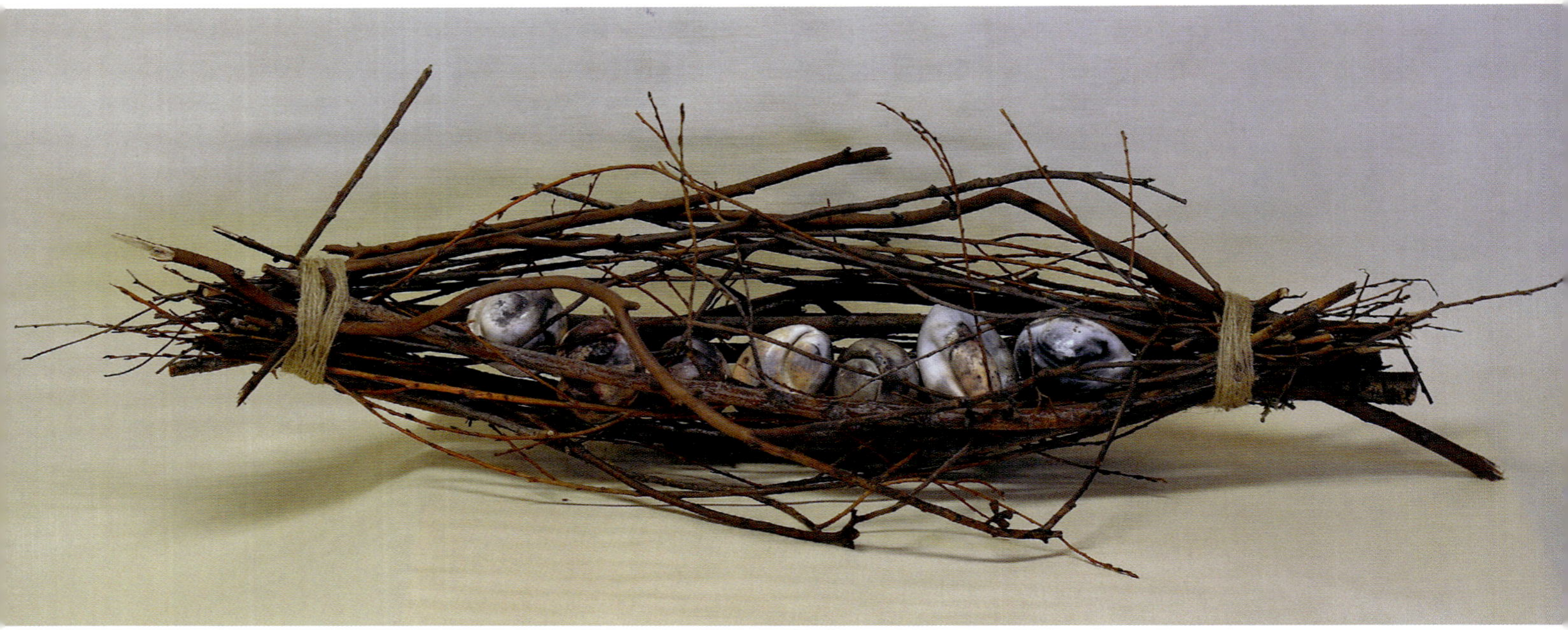

Dawn Whitehand. *Rescue Pod.* Wheel-thrown and manipulated, burnished, pit fired, waxed, eucalyptus, and wattle twigs, jute. 34" l x 12" w x 10" h. *Photo: Author*

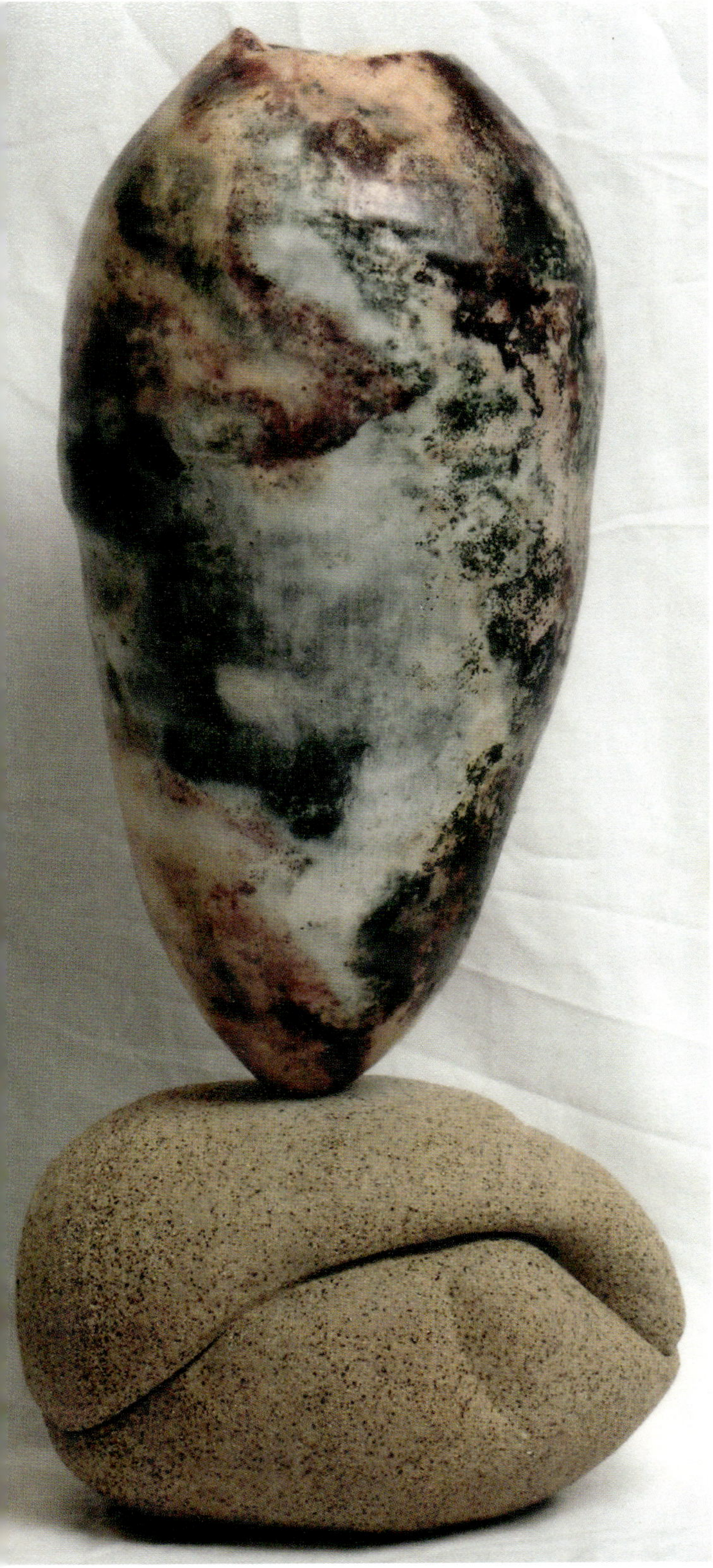

Dawn Whitehand. *Time Watcher.* Wheel-thrown and manipulated, burnished, pit fired, waxed, mounted on wheel-thrown and manipulated BRT form. 24" h x 11" w x 8" d. *Photo: Author*

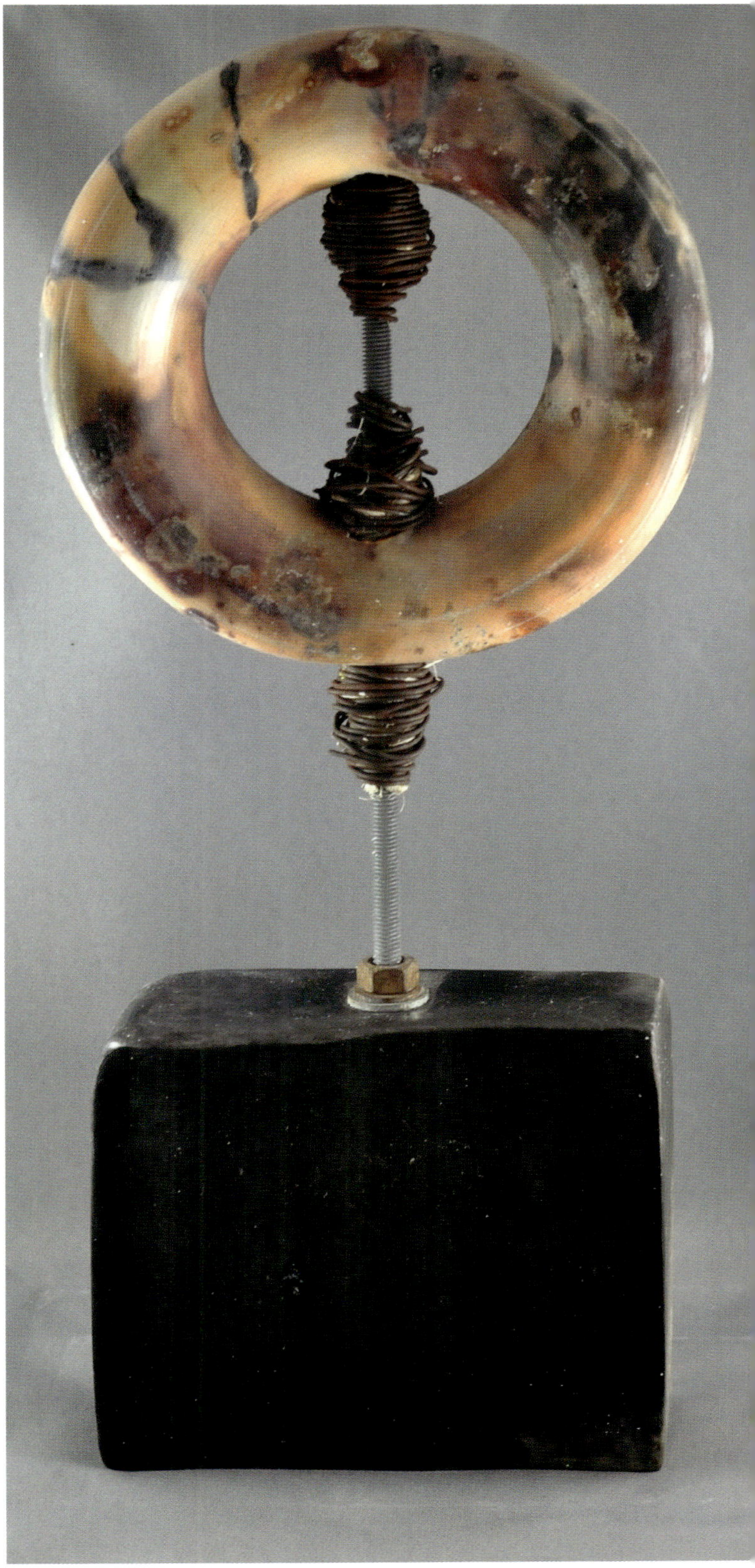

Dawn Whitehand. *Nature.* Wheel-thrown and handbuilt, burnished, pit fired, assembled using metal rod and rusty wire. 16" h x 6.25" w x 2.5" d. *Photo: Author*

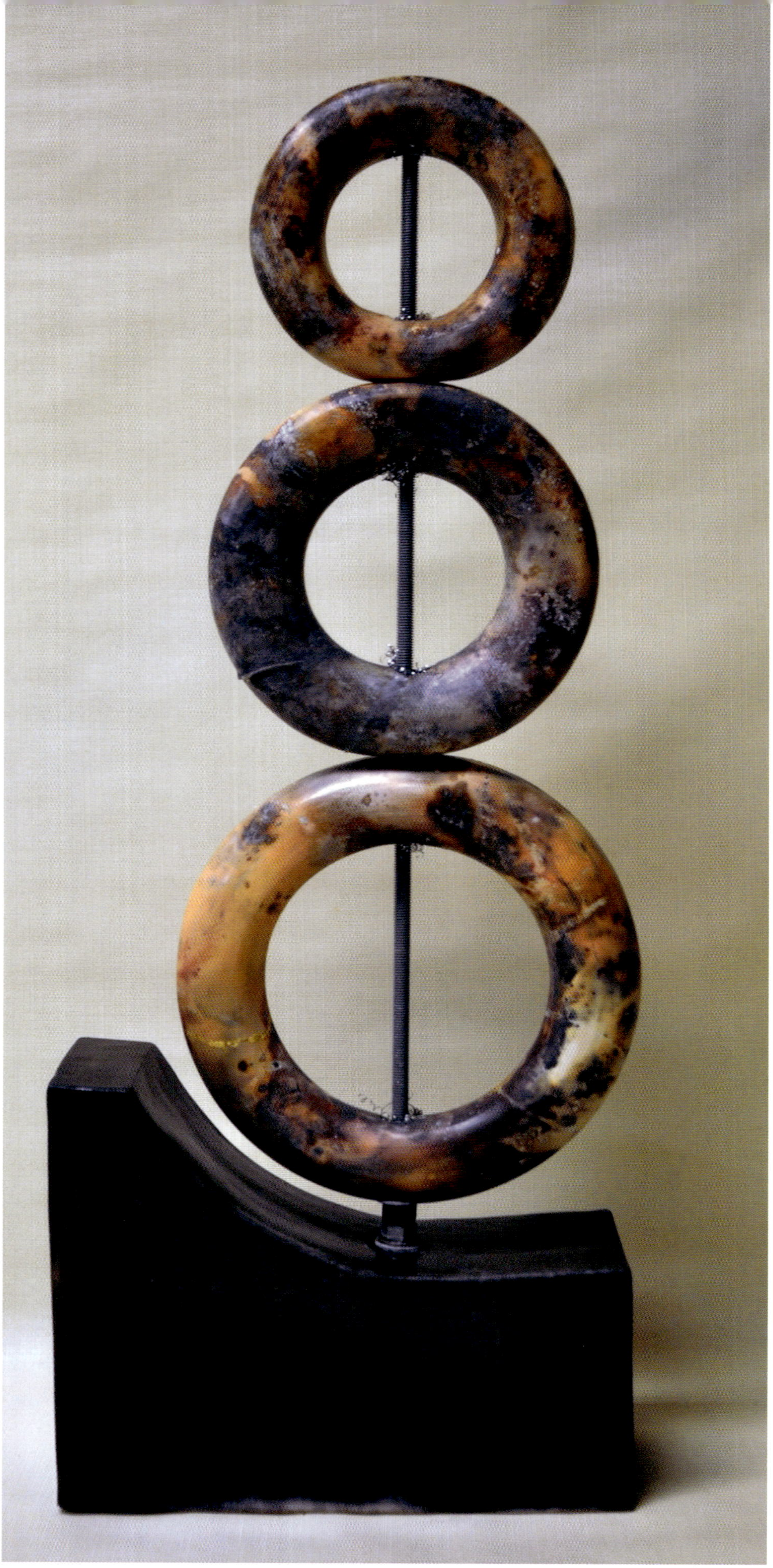

Dawn Whitehand. *Earthly Delights.* Wheel-thrown and handbuilt, burnished, pit fired, assembled using metal rod and metal scourer. 35" h x 12" w x 7" d. *Photo: Author*

CHAPTER FIVE

CONCLUSION

In the last analysis, the psychological roots of the crisis humanity is facing on a global scale seem to lie in the loss of the spiritual perspective. Since a harmonious experience of life requires, among other things, fulfillment of transcendental needs, a culture that has denied spirituality and has lost access to the transpersonal dimensions of existence is doomed to failure in all other avenues of its activities (Grof in Gablik 1991, 57-8).

In the context of twenty-first century concerns—global warming, over population, consumerism, globalization, technology, and the predictions forecast by writers such as Alvin Toffler in *Future Shock*—becoming a reality, pit firing is an attractive and highly individualized method within ceramic art practice. No matter how many pit firings are undertaken, the results will never be the same, and this unpredictability is what appeals to many ceramicists in an age of homogeneity.

While ceramics and the technological advancement of society are intrinsically linked, ironically it is the primitive, and somewhat uncontrollable, nature of pit firing that entices and excites the modern ceramicist.

The technology of modern times has enabled the use of ceramic equipment such as electric kilns and pyrometers, which have made the process of producing ceramics more efficient and reliable. While such technological developments may be advantageous for industry and the production of functional wares, many ceramicists crave creative spontaneity and chance within their artwork, and for this reason, pit firing as an alternative method of firing has become highly popular. Pit firing re-introduces excitement, the unpredictable, and vigor into the ceramic process and this energy is reflected in the dramatic surfaces that can be accomplished using this technique, eventuating in non-functional, expressive, and unique artworks.

The popularity of pit firing as a widely adopted ceramic method will also increase due to economical, environmental, and cultural issues. Economically, the cost of mass production and distribution is at a point where individual ceramicists producing functional ware cannot compete, and while many people still have an admiration for the "handmade," this is a small niche market: certainly not earning enough to make a "living" in most instances. The growing acceptance of clay as a fine art medium, and the dissolution of the age old art and craft debate, however, means that, culturally, ceramics is positioned on the verge of an artistic rebirth, and the non-functional, yet energetic product of pit firing dovetails ideally with this ceramics renaissance.

Many of the ceramic artists featured in this book choose pit firing as a studio method for the spontaneous and direct interaction it provides for the artist. And, as mentioned, the "unpredictable" factor is key. The artist succumbs to the elements, and is happy with what nature gives in return, just as our primitive ancestors gave ritual thanks to the Earth for what it provided. This connection with our historical forebears is also an attribute many of the featured artists mention, and it is this connection with history, a history where our ancestors had a much more intuitive and cyclic relationship with the Earth, that many social commentators claim technology has quashed in society (Whitehand 2009, 5).

As I wrote in my PhD hypothesis, it is this "disconnection" from the Earth that enables humanity to objectify our environment, viewing it as the "other," hence, until a connection is revitalized, and we become re-aware of our embedment in the universe, we will continue to exploit the Earth (Ibid, 5, 21, 111).

In an environmental context, therefore, pit firing can become emblematic: synonymous with sustainability and a need and willingness to return to a non-exploitative relationship with the Earth. Pit firing, therefore, can become instrumental in providing a conduit between humanity and the Earth in re-establishing and revitalizing this awareness, both for the makers and admirers of pit fired ceramics.

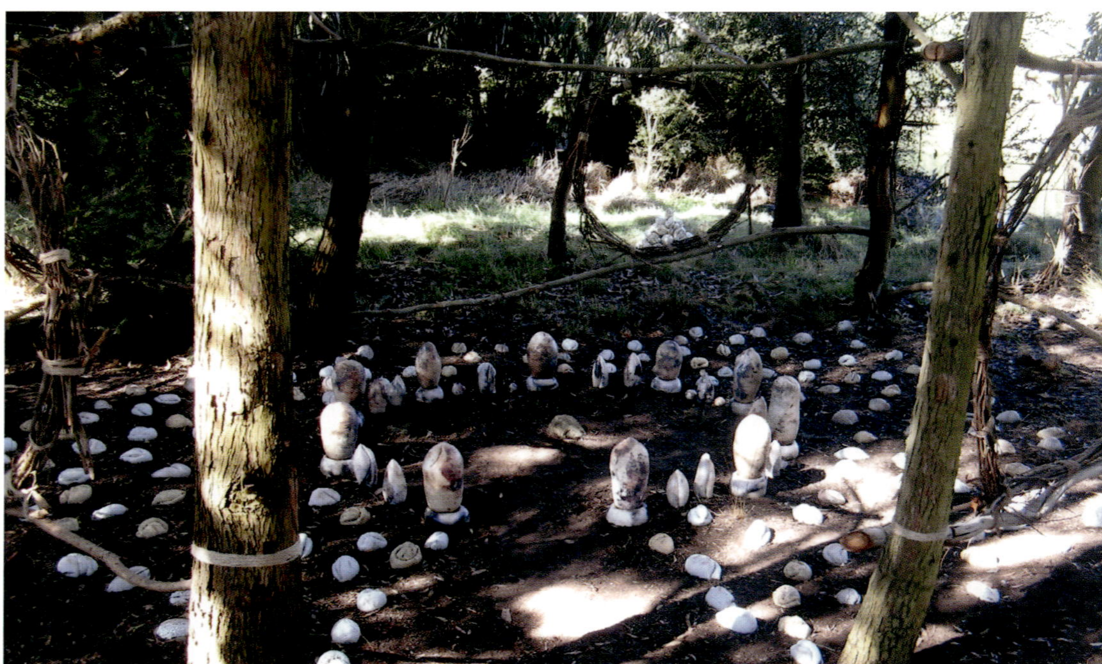

Dawn Whitehand. *Healing Circle.* Ceramic and mixed media environmental installation. *Photo: Author*

GLOSSARY

Bisque: Preliminary firing to ensure all moisture evaporates from the clay body. This firing must be performed slowly at initial stages to ensure steam can escape gradually, avoiding explosions. This firing also hardens the clay body for easier handling at the glazing stage.

Burnishing: Compresses and realigns the clay particles, contributing to a degree of waterproofing (utilized by tribal societies). It also achieves a sheen, which contemporary artists use for aesthetic purposes. Any smooth object, such as the back of a spoon or smooth stone, can be used for polishing. As the finished piece is still porous, due to low firing, it can be sealed with a wax product. This will also protect the delicate surface.

Centering: This process turns the rough clay ball into a smooth dome, showing no signs of oscillation as it spins on the wheelhead, enabling a successful, symmetrical form to be thrown.

Chuck: An open, thickly thrown piece of clay with a wide, thick base that is secured to the wheelhead with a thin coil of clay.

Greenware: Dry, unbisqued clay.

Grog: Crushed or ground-fired clay of various coarseness added to clay body to provide texture and strength, reduce shrinkage, open clay body, or provide thermal shock properties.

Incise: A form of decoration made by cutting/carving into the clay surface.

Leather-hard: Partially hardened clay due to evaporation. Clay is firm enough to handle without distorting, yet workable for turning, incising, clay additions, and slip application.

Oxidation: An oxidized atmosphere occurs when there is ample oxygen in the kiln. This will always occur in an electric kiln, and, in a gas kiln, occurs when primary and secondary air sources are left open during the entire firing. During traditional firings, such as a pit or open firing, oxygen is allowed to freely circulate by not smothering the pots in mud, sand, or thick shards.

Raku: A traditional Japanese firing method used to fire Japanese tea bowls. Introduced to the West by Shoji Hamada in the early 1970s, it was quickly adopted by progressive artists who wanted to push the boundaries of art and art's meaning. The method has been modified to include a post fire reduction where hot pots are removed directly from the kiln and are placed in organic materials and fumed for several minutes before being quenched in water to "freeze" the post reduction effects.

Reduction: A reduced atmosphere occurs in a kiln when the primary and secondary air sources flowing into the kiln are partially closed, causing the formation of excess carbon monoxide. To obtain oxygen, the flame leeches into the glaze ingredients and clay body, altering its chemical composition, and therefore changing the color of some glazes, such as copper reds. Forcing reduction in a gas kiln is more successful than in an electric kiln as the latter has no open flame. During traditional firings, such as a pit or open firing, pots can be deprived of oxygen by smothering the works with mud or sand, which causes a surplus of carbon monoxide and black smoke that carbonizes the pot surface.

Saggar: A container, traditionally made of clay and grog, to hold pots during a glaze firing to protect them from the dirty flame of oil or coal. Modern kilns generally don't use these dirty fuels anymore, so saggars are not needed. Instead they are used for the opposite: to isolate pots from the rest of the firing with organic materials within the saggar to achieve special effects.

Slip: Liquid clay about the consistency of single cream.

Slip casting: A method of making identical hollow shapes using plaster molds. Slip is poured into the mold and after a few minutes a firm layer has formed on the inside of the absorbent mold. After the desired thickness is achieved, the remaining liquid clay is poured out. The remaining clay shrinks away from the mold and the form can be removed.

Stoneware: Clay fired to a high temperature of between 2260–2370° F [1240–1300° C], giving it a stone-like appearance and strength.

Smoke firing: The smoke firing process is similar to pit firing, except that only sawdust is used to carbonize the clay surface. Coarse sawdust will allow for more oxygen flow, resulting in a patchy grey surface, while a finer sawdust will give a stronger, more even black surface.

Updraught kiln: A simple kiln operating on a chimney style principle. Fire is lit and fuel added at the bottom of the chamber and travels up through a grate on which pots are stacked. The heat is drawn through the ware by the opening at the top.

Wasters: Broken shards and discarded debris found around a kiln in an archaeological site.

Wheel-thrown: Term used to describe process of making forms on the potter's wheel.

BIBLIOGRAPHY

Amiran, Ruth. "The Beginnings of Pottery-Making in the Near East." In *Ceramics and Man*, edited by Frederick R. Matson. Chicago: Aldine Publishing Company, 1965.

Anonymous. Dictionary.Com [cited 24 July 2011]. Available from http://dictionary.reference.com/browse/neolithic.

Balfet, Hélène. "Ethnographical Observations in North Africa and Archeological Interpretation." In *Ceramics and Man*, edited by Frederick R. Matson. Chicago: Aldine Publishing Company, 1965.

Blinman, Eric, and Clint Swink. "Technology and Organization of Anasazi Trench Kilns." In *The Prehistory and History of Ceramic Kilns*, edited by Prudence Rice. USA: The American Ceramic Society, 1997.

Bow, Kim. "Lustred Ceramics." *Crafts Art International* 25 (1992).

Cooke, Glenn R. "Dictionary of Australian Artists Online" [cited May 14 2010]. Available from http://www.daao.org.au/main/read/7261.

Cooper, Emmanuel. *Ten Thousand Years of Pottery*. 4th ed: The British Museum Press, 2000.

Feinman, Gary. M, and Andrew Balkansky. "Ceramic Firing in Ancient and Modern Oaxaca." In *The Prehistory and History of Ceramic Kilns*, edited by Prudence Rice. USA: The American Ceramic Society, 1997.

Flood, Josephine. *The Original Australians*. Crows Nest: Allen and Unwin, 2006.

Francis, Keven. "Working the Pit." *The Journal of Australian Ceramics* 49, no. 3 (2010).

Gablik, Suzi. *The Re-Enchantment of Art*. New York: Thames and Hudson Inc., 1991.

Garraty, C.P. "The Origins of Pottery as a Practical Domestic Technology: Evidence from the Middle Queen Creek Area, Arizona." *Journal of Anthropological Archaeology,* March (2011).

Hoare, John, ed. *Clay Statements*. Toowoomba: Darling Downs Institute Press, 1987.

Koyama, Fujio. *The Heritage of Japanese Ceramics*. Trans. John Figgess. New York, Tokyo, Kyoto: Joint publication of Weatherhill and Tankosha, 1973.

Linné, S. "The Ethnologist and the American Indian Potter." In *Ceramics and Man,* edited by Frederick R. Matson. Chicago: Aldine Publishing Company, 1965.

Lockwood, Ken. "Playing with Fire." *Craft Arts*, 29-32 (1984).

Mellaart, James. "Anatolian Pottery as a Basis for Cultural Syntheses." In *Ceramics and Man*, edited by Frederick R. Matson. Chicago: Aldine Publishing Company, 1965.

Miller, Heather. "Pottery Firing Structures (Kilns) of the Indus Civilization During the Third Millennium BC." In *The Prehistory and History of Ceramic Kilns*, edited by Prudence Rice. USA: The American Ceramic Society, 1997.

Moore, A. M. T. "The Inception of Potting in Western Asia and Its Impact on Economy and Society." In *The Emergence of Pottery*, edited by William K. Barnett and John W. Hoopes. London: Smithsonian Institution Press, 1995.

Nelson, Glenn C. *Ceramics: A Potter's Handbook*. New York: Holt, Rinehart and Winston, Inc., 1971.

Perryman, Jane. *Traditional Pottery of India*. London: A & C Black (Publishers) Ltd., 2000.

Perryman, Jane. *Smoke Firing: Contemporary Artists and Approaches*. London: A & C Black (publishers) Limited, 2008.

Poole, Christopher A. "Prehispanic Kilns at Matacapan, Veracruz, Mexico." In *The Prehistory and History of Ceramic Kilns*, edited by Prudence Rice. USA: The American Ceramic Society, 1997.

Prier, Carol Molly. "Energy and Care: Pit Firing Burnished Pots on the Beach." In *Barrel, Pit and Saggar Firing*, edited by Sumi von Dassow. Ohio: The American Ceramic Society, 2001.

Rhodes, D. *Kilns: Designs, Construction and Operation*. Radnor, PA: Chilton Book Company, 1981.

Rice, Prudence. "Concluding Comments." In *The Prehistory and History of Ceramic Kilns*, edited by Prudence Rice. USA: The American Ceramic Society, 1997.

Rogers, Ray. "Pit Firing." *Pottery in Australia* 22, no. 2 (1983).

Rogers, Ray. "Pit Firing." *New Zealand Potter* 25, no. 2 (1983).

Rye, Owen. *Pottery Technology: Principles and Reconstruction*. Washington D.C.: Taraxacum, 1981.

Rye, Owen, and Clifford Evans. *Traditional Pottery Techniques of Pakistan*. Washington: Smithsonian Institution Press, 1976.

Scott, Lindsay. "Pottery." In *A History of Technology*, edited by Charles Singer, E.J. Holmyard, and A.R. Hall. Oxford: Oxford University Press, 1954.

Shimada, Izumi. "The Variability and Evolution of Prehispanic Kilns on the Peruvian Coast." In *The Prehistory and History of Ceramic Kilns*, edited by Prudence Rice. USA: The American Ceramic Society, 1997.

Sinopoli, Carla. *Approaches to Archaeological Ceramics*. New York: Plenum Press, 1991.

Vitelli, Karen, D. Franchthi. *Neolithic Pottery*. Vol. 1. Bloomington & Indianapolis: Indiana University Press, 1993.

Von Dassow, Sumi, ed. *Barrel, Pit and Saggar Firing*. Ohio: The American Ceramic Society, 2001.

Whitehand, Dawn. "Evoking the Sacred: The Artist as Shaman." http://independent.academia.edu/DawnWhitehand/Papers/534255/Evoking_the_Sacred_The_Artist_as_Shaman, 2009.

Dawn Whitehand. *Ancient Landscape*. Base—handbuilt, burnished, and pit fired in sawdust only. Vessel—wheel thrown and manipulated, burnished, and pit fired. 13" w x 4" d x 6 3/4" h.

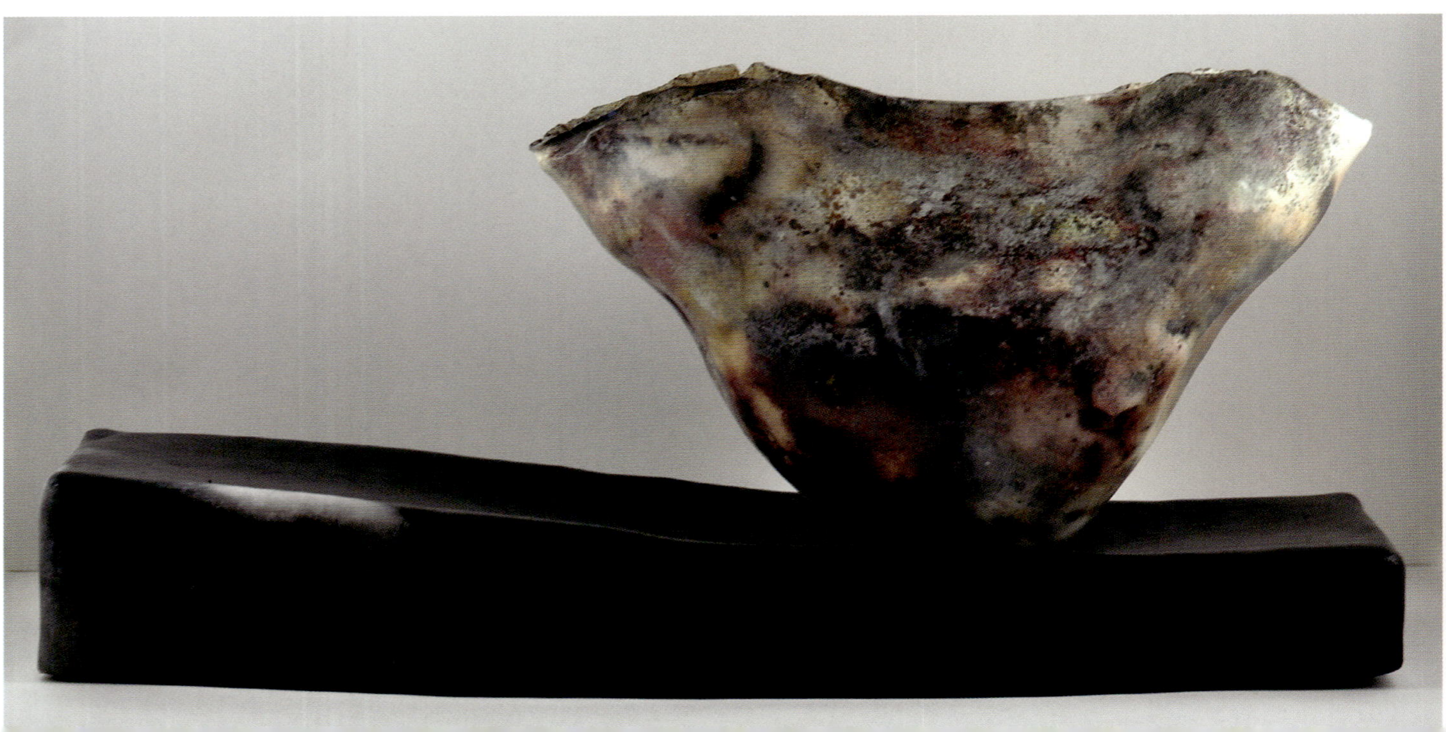

INDEX

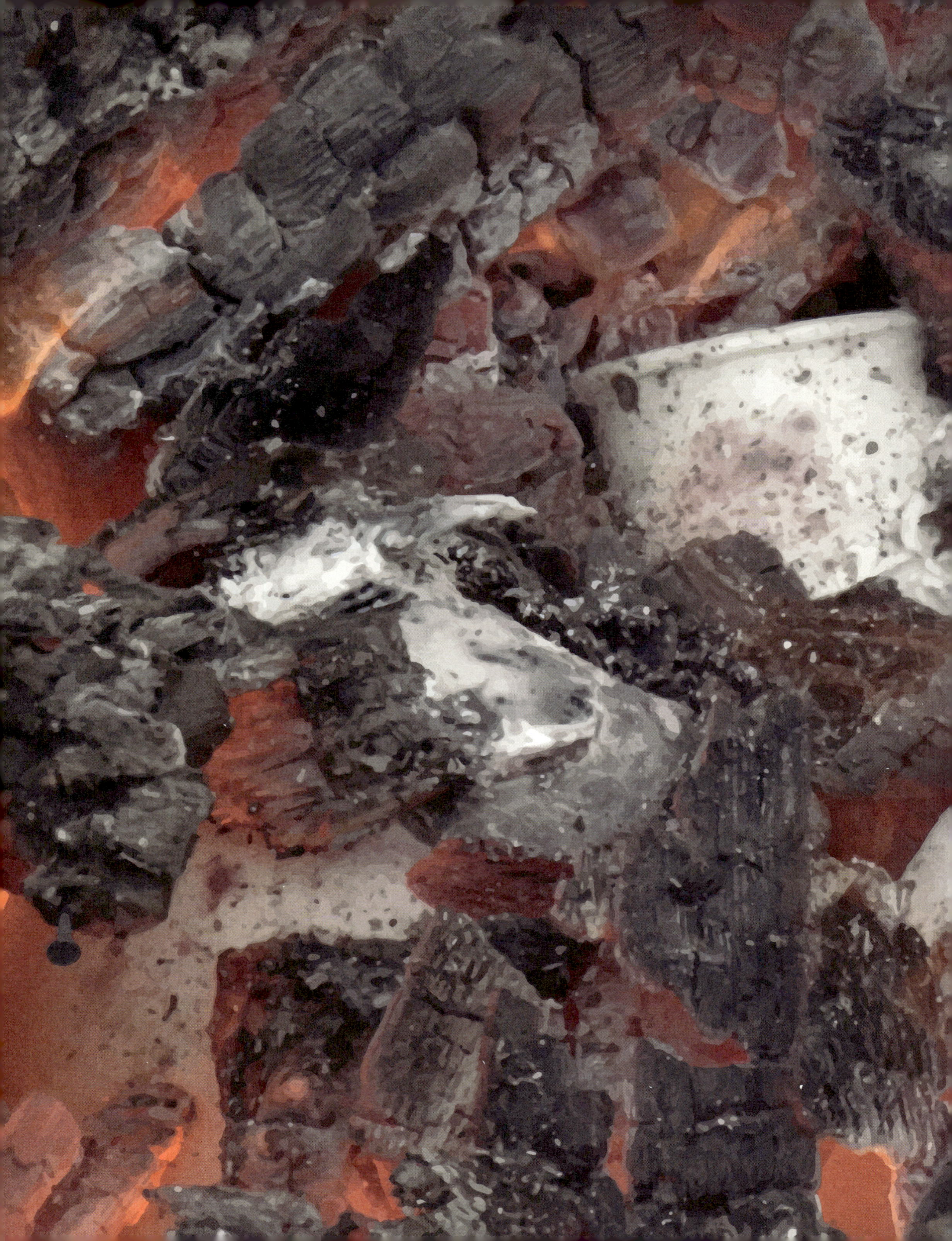